Community Networks

Lessons from Blacksburg, Virginia

Second Edition

For a listing of recent titles in the *Artech House Telecommunications Library,*
turn to the back of this book.

Community Networks

Lessons from Blacksburg, Virginia

Second Edition

Andrew Michael Cohill
Andrea L. Kavanaugh
Editors

Artech House
Boston • London

Library of Congress Cataloging-in-Publication Data
Community networks : lessons from Blacksburg, Virginia / Andrew
    Michael Cohill, Andrea L. Kavanaugh, editors.—2nd ed.
        p.   cm.—(Artech House telecommunications library)
    Includes bibliographical references and index.
    ISBN 1-58053-030-3 (alk. paper)
        1. Electronic villages (Computer networks)   2. Community life—
Technological innovations.   3. Blacksburg Electronic Village.
    4. Computer networks—Social aspects   I. Series.
    II. Cohill, Andrew Michael.   III. Kavanaugh, Andrea L.
        TK5105.83.C66   1999
        004.67'8'09755785 21—dc21   99-041777
                                                                     CIP

British Library Cataloguing in Publication Data
Community networks : lessons from Blacksburg, Virginia.—
    2nd ed.—(Artech House telecommunications library)
    1. Internet (Computer network)   2. Internet (Computer network)—
    Social aspects—Virginia—Blacksburg
    I. Cohill, Andrew Michael   II. Kavanaugh, Andrea L.
    004.6'78
    ISBN 1-58053-030-3

Cover design by Ariana C. Rork

© 2000 Artech House, Inc.
685 Canton Street
Norwood, MA 02062

International Standard Book Number: 1-58053-030-3
Library of Congress Catalog Card Number: 99-041777

10 9 8 7 6 5 4 3 2 1

Contents

Foreword

Robert C. Heterick, Jr.
Chairman of the Board,
Blacksburg Electronic Village, Inc.

A PEDDLER COMES INTO a village one day and says, "I can make delicious soup using only this stone—all I need is a big pot, some water, and a fire." Everyone in town comes out into the public square to scoff. Gradually, they are captivated by this fool and his rock. Why not give him what he asks for, they think—this could be entertaining. Having set up a pot over a fire, he drops his magic stone into the boiling water, saying, "This soup will be delicious, but a pinch of salt would improve it tremendously."

Willing to bend the rules to this minor extent, someone goes off to fetch some salt. By the same tactic, a little pepper is added, then a handful of herbs, a few carrots, several potatoes, perhaps a turnip or two. Meanwhile, kids are turning somersaults on the village common, dogs are barking, and a festive atmosphere is infecting the crowd. Everybody is chatting and gossiping and wondering how anyone could be dumb enough to think he could make soup from a stone. The concoction is beginning to smell pretty good, and our peddler remarks, "You know, this soup will taste wonderful exactly as it is, but a few chunks of meat would make it really excellent!"

A modern-day version of this story might have some itinerant network "techie" playing the role of the peddler, and the updated version of the stone soup would be a communications network bringing economic and cultural revitalization to the community. Such a story wouldn't be far from describing the beginning of the Blacksburg Electronic Village or any one of hundreds of similar experiments in community networking.

Only a very few folk with a vision for the future (a vision that probably is not very clear and not all that accurate) set out to convince a community that it (or the visionaries themselves) can build a community network like the stone soup of our fable—a few donated computers here, some communications links there. Most of those networks do not succeed, at least not very well, but a few develop a level of community acceptance beyond the vision of their founders. Fortunately for the tale of this book, the Blacksburg Electronic Village falls into the latter category.

Just exactly what is this vision that the peddlers of our updated story have in mind? And why should they be successful in convincing our 20th (almost 21st) century villagers that they can enjoy—and should contribute to—this new stone soup?

Observers frequently invoke a so-called wave theory to describe revolutions in how humankind relates to its surroundings. The first wave, the agricultural revolution, was spawned by the invention of the plow and some attention to fertilization, which produced a multiplier of about 100. The steam engine, a multiplier of 1,000, created the industrial revolution—the second wave. The computer and communications networks represent a leverage of 1,000,000 and have fueled the third wave—the information revolution. We had nearly half a millennium to absorb the agriculture revolution and something over a century to adapt to the industrial revolution. However, we will have not much more than a decade to understand how to deal with the information revolution. During the agricultural revolution, human labor was the driving force of society and land the measure of wealth. In the industrial revolution, capital was the driver and artifacts the measure of wealth. The information revolution will be different—knowledge will be the driving force of society and access the measure of success.

Community networking is about access to knowledge. Networking does not require the construction of edifices (although we may see a cellular or personal communication services (PCS) tower from time to time on the skyline), it does not involve manufacturing and the attendant problems of pollution, and it certainly does not need high population densities and the accompanying transportation gridlock. As Nicholas Negroponte so cogently observes in *Being Digital*, networking is about bits, not atoms. It is about information and the coding schemes used to represent the information, not the artifact in which information has historically come packaged. It is about moving information, not about moving people. It is about removing the constraints of time and place and lowering the threshold for society to access information and each other.

As delicious as this stone soup sounds, we should remember that there is a big stone in the bottom of the pot. A number of new and thorny problems will be raised as the glowing phosphors on our computer screens illuminate our schools, workplaces, and homes. Not the least of those problems will be the fundamental driving force—access.

There will always be a tradeoff between access and security. You can make something accessible or you can make it secure. We do not, and probably never will, know how to make information both accessible and secure. The problem is somewhat like that of finding a guard dog mean enough to scare off burglars but sufficiently docile so as to not bite the mail carrier (if we still have post offices in our electronic villages). This problem is playing out in the Congress even as I write this and will continue to do so for many years. How do we protect intellectual property on our networks? How do we ensure the integrity of financial transactions? How do we safeguard our national security interests? And how do we keep the pedophiles and pornographers at bay? Abraham Lincoln once observed, "There are simple answers to all questions—and most of them are wrong." These questions are not simple and will require clear thinking and consensus in our electronic villages.

In the world of bits, rather than atoms, intelligence and innovation will be distributed out to the nodes—every reader will also be a publisher; every listener, a composer; every viewer, a movie producer. Our classical hierarchical organizational models just do not work in the networked world. So from time to time, we will ask, "Who's in charge here?" And the surprising answer will be, "Us! We are in charge!" That is going to require a level of citizenship beyond that for which most of us have been educated.

The successful electronic villages will emerge from communities that are prepared to grapple with and are committed to resolving those kinds of difficult issues. The reward for winning such battles will be empowered citizens leading rich and full lives beyond any we have known in our history. Successful electronic villages will become magnets for economic development, quality educational systems, and satisfying cultural ambiance.

If our electronic villages are going to exhibit staying power, they will need to spawn new businesses and a true marketplace for electronic services. Our peddler is not going to stay in town forever. Those folks subsidizing our stone soup are not going to contribute their turnips forever either. There will need to be value offered for value received. We will need to decide whether, and how much of, the village treasury should be tapped for the village commons and how many and which commercial and government suppliers should supply the rest. Our history in that regard has not been the best.

We have tended to view our village infrastructure ambivalently, granting real or effective monopoly rights to some commercial providers (the phone and cable companies, for instance), using the taxing authority of the village for other infrastructure efforts (road, water, and sewer systems, for instance), while relying on village charity for yet others (fire and rescue services, for instance).

At the current stage of development of our digital world, it appears that strong competition leads to vigorous innovation, which in turn drives down prices, whether paid directly by the consumer or by way of the tax collector. One of the signals of success of the Blacksburg Electronic Village is the level of economic activity and competition that has developed for its services. As the federal government moves to remove the dead hand of regulation from our telecommunications industries, it would be a pity to see it reapplied by our village governments.

Competition is not the only cannon needed in the arsenal of our electronic villages. As the cartoon character Pogo was fond of saying,

"We have met the enemy and they are us." While strong competition for mature services drives down prices and generally increases quality, the nascent services are most effectively introduced through collaboration rather than competition. The Blacksburg Electronic Village signals successes in that area with its strong collaborations among Virginia Polytechnic Institute and State University (commonly referred to as Virginia Tech), Bell Atlantic, and the Town of Blacksburg. Many more collaborations are going to be necessary to sustain and develop the electronic village.

Lest I be accused of flying under false colors, let me confess that I was a member of the faculty of Virginia Tech for 30 years, Bell Atlantic is my phone service supplier, I pay taxes to the Town of Blacksburg, and, last but not least, I am an enthusiastic subscriber to the services of the Blacksburg Electronic Village. I might add that I maintain all those relationships quite happily.

Preface to the First Edition

by Vinton G. Cerf
Senior vice president,
MCI Communications Corp.

THIS IS A STORY about the real world and the real-world effects of the ethereal cyberworld. Networking is about *more* than bits. The network and its parts must be made manifest in computers and other devices that are a part of the physical world. Homes need reorganizing, as do businesses and other institutions, in order to accommodate network-enabled devices. Physical communication facilities are needed, whether cellular, fiber, copper pair, coaxial, or satellite based. Networks are not divorced from the real world. They exist in it and are anchored to it, together with the people who use them.

Many actions on the Internet have real-world consequences. Not all of this activity is so directed, but much of the Internet's value is found in its real-world effects. This does not detract from the creative ferment of the digital world. We still paint on an endless electronic canvas with an unlimited software palette, but we should not forget the many real-world bindings that our digital paintings enjoy and enhance. It is the meeting of these two worlds—the real and the "ether-real"—that gives networking, and the Internet in particular, its great power.

Andrew Cohill is right to characterize the Internet as a *new medium*. As such, it is not without its problems. It has finite capacity, and its parts sometimes become overloaded, especially in this period of tremendous growth. It is also possible for its users to become overloaded with information, such as e-mail, responses from web-searching tools, and applications that consume more time than anticipated.

The Blacksburg experience is important for many reasons, not the least of which is that it offers real-world experience for the networking phenomenon. Blacksburg is a technological microcosm that has been and will continue to be chronicled in some detail by those involved in its creation and who live day to day in its evolving universe. The Internet community as a whole can learn and benefit from the successful and unsuccessful applications that have been tested and are being tested in Blacksburg.

The Blacksburg experience marks an important milestone in the evolution of the Internet—I await with great anticipation the next installments in the future history of this online community.

Preface to the Second Edition

by Andrew Michael Cohill, Ph.D.
Director
Blacksburg Electronic Village

A S I WRITE THIS NOTE, the Blacksburg Electronic Village is about to complete six years of serving the citizens of Blacksburg and southwestern Virginia. It has been nearly three years since I began work on the first edition of this book. In that time, much has changed. Today, most of the citizens of Blacksburg are online. At the last couple of civic meetings I attended, when the question was posed, "How many folks have e-mail?" everyone in the room raised their hand. Nationally, estimates vary about what percentage of the country is online, but the most optimistic suggest that only about 35% of U.S. citizens have Internet access. Blacksburg is truly a "connected" community.

With most of the world still waiting for affordable telecommunications, much work remains to be done. Worldwide, there are community network projects on every continent except Antarctica. Rural and disadvantaged communities must begin investing time and energy in community-wide networks (public and private) to ensure that they are not left behind in this new information economy. The development of public water systems providing clean water to citizens and communities has been a major focus of community effort in the 20th century. In the 21st century, communities need a new kind of piping system— data pipes, connected to every home, every business, and every public instititution.

Another apt analogy is road building. We began the 20th century with few paved streets. We finish the 20th century with national and international road systems, designed and managed for the public good and built and maintained by the private sector. Communities need data highways that have been designed to serve the entire community affordably. Can anyone imagine a situation in which a single company (or perhaps just two) own and control all of the roads in a community? Where toll booths could be erected and fees charged just to drive downtown to the grocery store?

Unfortunately, many communities have that situation now with respect to data highways. Too often, a single corporation with managers hundreds of miles away are making decisions that affect community futures, especially in economic development. Clean water and good roads and highways have been the cornerstone of community economic development in the 20th century. In the 21st century, communities need high bandwidth data pipes and affordable information highways connected to and within the community.

The good news is that data pipes do not leak, and information highways do not wear out at the same rate as asphalt roads. Even very small communities can build a modern telecommunications infrastructure. In Blacksburg, we are trying to learn the best way to do this, to connect every community to the world.

1

Welcome to Blacksburg

by Andrew Michael Cohill

JOE, AN ELDERLY MAN in Blacksburg,[1] suffered a fall three winters ago and broke his hip. Joe had to remain in bed for almost two months while his hip healed, but he still spent a lot of time with his friends. Joe is a member of the Blacksburg Electronic Village (BEV) Seniors group, which represents one of the most cohesive segments of the community in terms of online activities. With his personal computer at his bedside, Joe stayed in daily contact with his friends by using the BEV Seniors mailing list to keep up with group discussions and e-mail to maintain personal contact with friends and neighbors. The asynchronous nature of these communication tools means that Joe could keep up with his mail during the day when most of his friends were busy with other activities, and that his friends could reply to Joe at their convenience. Joe healed quickly and was soon back on his feet. The daily contact and the rich communications potential of electronically written letters kept

---

1. All the stories described here are true; some names and places have been changed to protect the privacy of the individuals involved.

Joe occupied and alert throughout his convalescence and allowed a group of busy friends to support him in a way that would not have been possible without Joe's connection to the Internet.

Ellen, a woman with a strong interest in natural foods, has always wanted to run her own natural foods business. It never seemed possible, given the high costs of renting space, hiring employees, and stocking and managing inventory. Today, Ellen runs her natural foods business out of her tiny one-bedroom apartment. She manufactures specialty vinegars, offering a wide variety of vinegars flavored with organic herbs and spices. Ellen does not need a storefront because she advertises and sells her products via her Web site, which provides her with a small but growing national market for her products.

Lynn, a single mother with a teenage son, was working as a part-time meter reader two years ago. Anxious to increase her income and find better work, Lynn began studying Web design. She was able to work in her apartment because she had affordable Internet access in Blacksburg, and she is able to manage her time around her son's school schedule. Today Lynn is making more than three times her previous salary as a meter reader and has a small but growing clientele that relies on her expertise as an Internet consultant and Web designer.

Matthew is a retiree living in Blacksburg whose grown daughter lives in another state. She has been diagnosed with a serious disease but does not feel that she is getting adequate treatment from her local doctor. Matthew is frustrated by the situation and turns to the Internet for help, even though he does not have a computer at home.

He visits the public library in Blacksburg and spends hours online, learning how to use Internet information tools like gopher, mailing lists, search engines, and the World Wide Web. After reading and studying online resources, he finds a mailing list where people discuss the disease that afflicts his daughter. On that list is a doctor who specializes in treating her disease. Using e-mail and then the phone, Matthew contacts the doctor and arranges for his daughter to travel to see the doctor for a diagnosis. The doctor is able to prescribe a new medicine she had not taken before, and her condition improves.

In the public school, two sixth grade teachers have dreamed for several years of engaging their students in the design, writing, and production of a history of the area, based on the stories and reminisces of family and friends. They have always put off the project because the effort and expense of doing it on paper seemed to outweigh what the students might learn from the experience. They had talked about trying the project in an electronic format, but were not sure what the best way to do it was.

Fortunately, a graduate student from Virginia Tech who was looking for a research project was introduced to them, and when they described what they wanted to do, he showed them the World Wide Web. In a short period of time, he had demonstrated how easy it was to publish information on the Web. The teachers finally were able to try the experiment, and within a couple of months a group of sixth graders were excitedly writing, editing, and publishing their local history for a county-wide and worldwide audience. Seeing the results of their writing and editing online in a few hours or a few days excited the children. It made writing come alive for them in a way that traditional English composition had never done.

Just hook people up

If we have been successful in the BEV, it is because of one simple idea: Just hook people up however you can. What matters is not how people are connected, but if they are connected. For many years to come, there will always be cheaper, faster ways to hook people up just around the corner. For a long time to come, it may seem less risky to wait for tomorrow's technology rather than to bet on today's technology, which may be obsolete in three years. We have seen this phenomenon when it comes to purchasing a personal computer, too; something better is always just around the corner.

A community network is not about technology, it is about giving people a new way to communicate with each other. Depending on the kinds of connections you have in your community, users of the network will do different things with it, but e-mail, mailing lists, newsgroups, and forums all work at any speed and even with older computers. There is no reason to wait for any company or for any kind of promised network service. Dialup access to the Internet using ordinary phone lines and modems is now widely available, and modem access is all you need to start a community network in your town. In many areas, there are now faster alternatives as well, like cable modems, ISDN, and xDSL services. Each service creates more competition, lowers prices, and offers individuals more choice.

In Blacksburg, with more than 85% of the community online in 1999, the goal of getting people connected has been accomplished. However, much work still remains. We are just now beginning to see the real value of a connected community; civic groups, with the entire membership online, are abandoning paper newsletters (often their single biggest cost) and are now providing more timely information via mailing lists and Web sites. At the same time, these groups are seeing their membership rise, and attendance at meetings has improved.

Neighborhoods are using mailing lists and Web sites to develop support for preserving and maintaining the quality of the neighborhood. In Blacksburg recently a neighborhood group successfully used a mailing list to alert neighbors to a potential zoning change that might have had a harmful impact on the neighborhood. The mailing list saved time because fewer meetings were needed but everyone had the facts they needed to make a decision. Names and addresses of planning commission members and town council members were distributed quickly and easily.

Businesses that have recognized the Internet as a strategic market for their business have successfully integrated networking into their daily business and rely heavily on the Internet to keep in touch with customers and to provide better, more timely service. Many of the newer, high-tech startups in Blacksburg say that they could not function as a business without affordable, high speed access—the Internet has become that critical.

The BEV is intensely personal

In all of human history, there is no precedent for what is taking place in Blacksburg, Virginia, and other communities. It is now possible for a single person, sitting in his or her own home, to broadcast a message to hundreds, thousands, even tens of thousands of people without any inter-mediaries and at very low cost. In the past, individuals with a message had to rely on others to broadcast that message, using books, news-papers, radio, or television. The cost was high because all the old media required substantial capital investment in equipment to begin disseminating information. Costs remained high even after the equipment was amortized because of limitations of the media (only so much information will fit in a daily paper or on a 30-minute news broadcast).

The Internet has the potential to grow in capacity nearly without limit, and the cost of distributing information is so low (a computer and network access) that many millions of people are already deeply involved in writing, talking, publishing, and exchanging ideas on the network. Powerful search engines index all this information day and night, so that it is possible to find information about almost anything with just a few keystrokes in just a few seconds.

Search engines have changed the way we all think about and use information. Information tools like Lycos, Yahoo!, Alta Vista, and MetaCrawler make it easy to locate the information we need quickly. It is no longer necessary to make copies of certain kinds of information and store it locally—the "Net" never sleeps, and hundreds of thousands of servers all over the world provide information day and night in dozens of languages.

Rote learning is diminishing in importance. In the past, what was important was how much we memorized and stored in our heads. In the future, different kinds of knowledge will be more important:

- How well we are able to find what we need to know at any instant;

- How well we can take all that information and turn it into useful ideas.

It is not the information itself, but how we use it that makes us smart.

In the past, models for teaching and learning have been based on the idea that information is scarce. That is no longer true. Information is cheap and getting cheaper by the moment, partly because it has been decoupled from more expensive media like paper, television, and videotape and partly because more people are now able to produce and publish it.

In Blacksburg, most of us are not wearing virtual reality goggles; most of us are not sitting in our basements in the dark, staring at computer screens; most of us have not changed in any substantial or noticeable way. We all just communicate with each other more easily and more conveniently. What the network really buys us is time. We complain about how little time we have; having this new communications channel in our homes simplifies routine communications tasks and makes it easier to keep in touch with people in the community who are important to us.

In Blacksburg, it is now easier to find out the schedule of the Swing Dance Society because the group has a home page (http://www.bev.net/community/swingdance/). It is easier to find out when classic movies are playing at the refurbished Lyric Theatre because there is a home page schedule (http://www.mfrl.org/compages/lyric/) for it as well. The BEV Seniors group (http://www.bev.net/community/seniors/) finds it easier to notify members about physical meetings because they all have e-mail (phone trees are obsolete).

We write more than ever in Blacksburg

The interesting thing about being connected is how much more time you spend writing than ever before. In Blacksburg, the most common Internet activity is not Web surfing, but e-mail, and e-mail is all about writing. Other common services in Blacksburg include mailing lists (writing and listening) and newsgroups and forums (writing and listening). Users of those services experience a rich two-way interactivity with other people (the World Wide Web is still essentially a one-way medium).

This new communications channel in town offers one-to-one conversations (e-mail), one-to-many conversations (mailing lists), and many-to-many conversations (newsgroups). The two most common topics of discussion online in Blacksburg are where to get your car fixed and where to eat in town. The car repairs discussion group is a valuable source of information about where all the good (and bad) garages are in town; folks can drop in anytime they have a question about their car or need help with a problem.

Food discussions often last a week and frequently involve 50 or more people at a time. Individuals talk about their favorite foods, their favorite restaurants, and whether or not the Harris Teeter grocery store has lower prices than the competition.

Some people fear that using the computer will diminish human contact, but just the opposite appears to be true in Blacksburg. Many things in our society have diminished our sense of community, but the network is not one of them. Suburbanization of our cities, which began after World War II, has been a chief culprit. As we moved out into the suburbs, we became increasingly isolated from our neighbors and from the community at large, because the suburbs segregated us from community gathering places like the corner store, the mom and pop grocery, and downtown areas.

The reliance on the automobile for travel instead of mass transit and our feet reduced the incidences of chance encounters with friends and neighbors. In Blacksburg, however, as in many other places around the country now, we can drop by a community gathering place on the Net 24 hours a day, seven days a week, to chat, catch up on local events, get help with a problem, or just hang out.

The Net is neither the death of physical place nor a replacement for physical place. But it can and does complement physical place and give us a new opportunity to communicate with each other in a public space and to talk with the entire community about the things that interest or concern us.

Storytelling

Storytelling is a human tradition that reaches back thousands of years. Human beings are different from animals in many ways, but one important difference is our love of stories and storytelling. In Blacksburg, each person can tell his or her own story free of intermediaries. This is what makes the BEV so powerful; we no longer have to ask (or beg) permission to tell others what we know or who we are.

Children in Blacksburg schools routinely publish their schoolwork on the Net for the whole world to see. I cannot help but think that for children growing up in isolated rural areas, this will change the way they think about themselves and about their place in the world. These same school children regularly converse with pen pals all over the world. The immediacy and freshness of these conversations motivate and excite the children to do well, to tell their stories well; they have a sense of power and accomplishment. The jury is still out on the most effective way to use technology in the classroom, but every piece of e-mail the children send is a thought put to words, a sentence composed, a letter written, another human being touched by their thoughts.

Television is a kind of storytelling, but it is a one-way medium that engages only the teller, not the listener. Storytelling is a communal activity that requires both a teller and a listener, but television tells its stories whether or not anyone is listening, and that eliminates any possibility of a meaningful dialogue. We killed writing with our lifestyle, and we as a country have made a long series of decisions that have taken us far down the road to personal isolation and a feeling of lost community. Community networks will not automatically fix all the problems created by those feelings, but it is a tool to start reconnecting ourselves to others. It is a tool to start telling others about ourselves. It is a tool to start listening to others again, to engage once again in communal storytelling.

Sense and sensibility

The role of government concerning the Internet has been the subject of vigorous debate. I will not attempt to summarize that debate here, but it seems appropriate to quote Thomas Jefferson: "The government that governs best is the one that governs least." Congress may be the greatest deliberative body in the world, but the Internet is not likely to wait for Congress to deliberate, now or in the future. The issue of decency is a perfect example of how sincere people with a complete lack of understanding of the technology are trying to regulate something that is completely intangible and untouchable.

It has been said so often that it has become a truism, but it is still true: The Net regards censorship as damage and routes around it. For the most part, our legislators have failed to understand that the technology of the Internet simply cannot be regulated the way things in the physical world are regulated. Government can set speed limits and enforce them in the real world because we have few choices if we want to drive from point A to point B; the cop with the radar gun simply goes where the traffic is. On the Internet, it is possible to take another route whenever we wish; there are many roads, and they can take us wherever we like.

The ethereal nature of information on the Net makes things like regulation of adult content equally pointless. Outlaw dirty pictures, and soon you will find people swapping pictures of Kathy Lee Gifford that have a dirty picture encrypted inside. Only people with the right encryption key will be able to find the dirty picture. Everyone else will simply see a chaste picture of Kathy Lee.

Decency on the Net is an important issue, but it can and should be dealt with locally whenever possible, preferably in the home. Parents can and should manage their children's use of the Internet, and there are some very simple solutions. If the family computer is kept in the den or the living room, your 13-year-old son with overactive hormones is not likely to spend much time downloading nude pictures of Cindy Crawford.[2] A computer that is located in a public area of the home can become a gathering place and tool that enables family members to connect, literally and figuratively, to the rest of the community.

Adolescent boys used to learn about sex out behind the garage with pilfered *Playboy* magazines. Sex on the Net is no more graphic than in our society in general: just turn on the TV and observe the language and images on shows that appear later in the evening. The network brings our society right into our own homes, and we do not like what we see. The problem begins when we decide to shoot the messenger rather than turning the rather bright and unpleasant spotlight on ourselves.

The solution is not control but personal responsibility. The chat rooms that Ann Landers claims are destroying marriages are simply a tool for communicating in a healthy or unhealthy way. Chat rooms do not make people give up on their marriages any more than beer does; if someone is drawn to a chat room (or alcohol) in an unhealthy way, the problem is with that person, not with the chat room (or the beer). The chat room and the beer may appear to aggravate the problem, but no one advocates giving up cars because people die in automobile crashes daily.

There is a legitimate role for government on the Internet. The Clinton administration deserves some credit for making government information and services more accessible by pushing all the major federal

---

2. Because of the inexplicable popularity of Cindy Crawford pictures on the Net, I have proposed a new measure of network performance called the Crawford, which would be how long it takes to download a 1-megabyte picture of Cindy Crawford.

agencies to publish their information on the Net. Not only does the Internal Revenue Service have a Web site that makes every IRS form available online, it also tries to be lighthearted at the same time, an intriguing change in the way we think about that government agency. Other possible government roles include the electronic equivalent of postmarks and providing secure repositories for online legal documents.

Education

With respect to education, we seem to have a national defect of assuming that bigger is better, what I call the atomic bomb approach to solving problems. Americans do not like to take the time to focus on the real problem and then follow through. We prefer to fly over the problem at 50,000 feet and drop an atomic bomb on it. That way, we do not really have to meet or talk to the people below whose lives will be inevitably changed; we can fly home quickly without any muss or fuss. Often, however, our national problems need the equivalent of a long messy ground war, with thousands of foot soldiers thrown into the fray, a solution we avoid.

A school district in the Pittsburgh area, over a three-year period, put a network connection in every school and a computer in every classroom simply by reallocating 2% of the school budget and going cold turkey on paper once there was a computer in every teacher's classroom. E-mail and Web pages were used for all internal communications.

In education, money is never the issue when it comes to connectivity; the issue is about power and turf and fear. Anyone who says there is not enough money in the school budget for computers and network connections simply does not want to face the inevitable shifts in power that the network brings. (The one exception may be some poor rural and inner-city schools that are shortchanged in budget allocations because of how state funds are distributed.)

One of the most unfortunate unintended consequences of network and computer technology is the often justified backlash against the technology by teachers who have been guinea pigs for the past 20 years in numerous failed technology breakthroughs that rarely solved any problems and often created new ones. Technology in the classroom will not solve existing education problems like underpaid and overworked teachers, severe student discipline and behavioral problems, or bloated administrative bureaucracies.

Regrettably, many technocrats promise that putting computers in the classroom will change education. Technology will change education most dramatically when teachers are paid properly, when parents get involved with their children's work, and when teachers get the time and support they need to adjust the curriculum to use the new network-enabled tools and resources.

There are many ways to see that these changes happen, but here are a few possibilities:

- Make grants directly available to teachers to buy their own computers in return for content development that adheres to a simple framework that facilitates reuse and wide distribution.

- Give teachers grants and salary supplements to create content that is network-ready.

- Make grants available to teachers for training out of the classroom and away from school. Sending teachers away for a week to learn about the network is the quickest and least expensive way to ensure wide use of the network in the classroom.

- Just pay teachers more.

- Give equipment grants to schools and school districts that agree to allocate existing budget dollars to provide technical support and training for teachers.

- Ignore the schools completely and train the parents in the language, use, and culture of the Internet. Parents ultimately pay the taxes to support the schools and often elect the school boards that run the schools, so parents who are network-literate will apply the needed pressure to see that the schools make the necessary changes.

Children and their parents do have the political power to change our schools, train our teachers, and reform our school administrators. Ultimately, this is a community responsibility that starts with the parents, but ideally it should involve every segment of the community, including businesspeople, who stand to gain from better educated workers; politicians, who want to stay in office; and school administrators, who should recognize that change is inevitable.

One of the best reasons for a community network is heightened awareness in the community of the value of being connected. Parents are better able to understand how teachers and students can benefit from the network when they themselves are connected, and they are more likely to provide the support needed for changes when they understand what it is they must ask for.

Blacksburg and the Net

This book tries to describe what has happened in Blacksburg as people have become connected in this new way and how we became connected. Our experiences in Blacksburg are not the experiences of an inner-city community struggling to find the resources to get connected or the experiences of a Native American community network project in New Mexico, but perhaps what we have learned will help others understand better what can happen.

Any community that wants to be connected can be. Size and money are unimportant; what matters is the only thing that has ever been important in matters of community: a few dedicated people with an idea and determination.

In 1993, when we began offering Internet access to the Blacksburg community, about 200 people were using the Internet in town. At the time, the notion of everyone owning a computer and using it to be "wired" seemed nothing short of peculiar to most folks. Today, about a third of the country is online (where Blacksburg was in 1995), and URLs and e-mail addresses are tacked onto everything we see.

The Internet has arrived. It is the fastest growing technology in the history of the world. It has changed us; it is changing us still; and it will continue to change us long into the future. Technology is not the problem. Lack of community is the problem, and education is the solution. We must find better ways to preserve what is good in our communities, and we must find better ways to sustain our local systems (education, public services, local government, civic institutions). We are, perhaps, in year 10 of a 50-year change. It will take that long for this new technology to mature, and it will take that long for us to change and adapt to this new technology.

In Blacksburg, we are learning who our neighbors are because we have this new technology, this new way to communicate. Technology will not solve our planning problems, but working together, as neighbors, we can meet any challenge.

About the book

Chapter 2: "A Brief History of the Project" was written by three of the people who helped start the effort and begins with the genesis of the project in the mid 1980s, long before the Internet was a household name. This chapter traces the development and design of the project from its inception to the present and discusses the roles played by many community partners in the effort.

Chapter 3: "The Architecture of a Community Network" was written by Andrew Michael Cohill, the current director of the project, and contains a high-level overview of the components of a community network. The major parts of the network specifically designed to support users in Blacksburg are also described in Chapter 3.

In Chapter 4: "Evaluating the BEV," Scott Patterson, a researcher at Virginia Tech, provides a summary of several research studies performed before the official start of service and during the first two years of its use.

Chapter 5: "Measuring the Community," is an extensive description of the research conducted by the Blacksburg Electronic Village over the past four years. One of the key findings is that people report feeling more connected to the community after going online.

Chapter 6: "Community Dynamics and the BEV Seniors Group," by Federico Casalegno, is a fascinating look at some of the changes that have taken place in the seniors community in Blacksburg as a result of the BEV.

Chapter 7: "Networking Families Into the Schools" by Ehrich, Lisanti, and McCreary describes a visionary three-year experiment to turn a fifth grade classroom in a tiny rural school near Blacksburg into the classroom of the future and the surprising effects on teaching, the students, and the family.

Chapter 8: "Managing the Evolution of a Virtual School," written by Roger Ehrich and Andrea Kavanaugh, two Tech research scientists, discusses the many challenges confronted by K–12 schools as they try to incorporate Internet use into the daily activities of the classroom. Choices and challenges faced by the local school district are covered in detail.

Chapter 9: "Learning and Teaching in a Virtual School" is a personal account of using the Internet in the classroom by Melissa Matusevich, a Blacksburg teacher with many years of experience with computers and the network.

Chapter 10: "Community Network Technology" by Luke Ward, the BEV technology manager, is written for readers without a technical background who want to know more about how a community network is put together. Much detail is provided about various ways to provide network services to individual users, schools, libraries, and businesses.

Chapter 11: "Managing Information in a Community Network" by Cortney Martin and Andrew Cohill describes the evolution of the BEV information services. The success of any community network depends on the active participation of people. This chapter discusses the strategies used to help people become accustomed to publishing and using online information.

Chapter 12: "Building an Online History Database" by Will Schmidt and Andrew Cohill is the story of the development of a Web-based information tool designed to record the history of an entire community, where everyone in that community is connected and can be an author who contributes either personal or communal accounts of local events.

Chapter 13: "Success Factors of the Blacksburg Electronic Village" was written by Andrew Cohill, the current director of the project. It describes some of the key events and activities that led to the very high use of the community network in Blacksburg.

Chapter 14: "The Future of Community Networks" by Andrew Cohill, discusses the challenges communities face in the 21st century as they integrate technology into the fabric of the community.

2

A Brief History of the Blacksburg Electronic Village

by Philip (Theta) Bowden, Earving Blythe, and Andrew Cohill

Getting started

As early as 1979, a number of faculty and graduate students in Virginia Tech's Urban Affairs and Urban Planning programs were interested in the concept of community-focused, network-accessible information products and services. Those ideas coalesced around the concept of a "community information utility." Throughout the 1980s, these individuals brainstormed issues concerning the appropriate technological and institutional form of—and market and regulatory obstacles to—such a utility.

Virginia Tech has always been a major influence in the life of Blacksburg, telecommuting (using the network from home to do work) being a good example. Beginning in the mid 1980s with the arrival of a new telephone system on campus, some faculty and staff began lobbying the university to provide better off-campus access to the mainframe computer.

The new phone system provided voice and data connectivity at every desktop; each phone had the equivalent of a 19,200-baud modem built into it. That was in 1987, when a 1,200-baud modem was considered state-of-the-art. Interest from university members who lived in Blacksburg and from the local phone company led to the creation of the BEV.

Three major entities combined to form a core public-private alliance to bring the BEV into being: Bell Atlantic of Virginia (formerly Chesapeake and Potomac Telephone of Virginia), Virginia Tech, and the Town of Blacksburg. This alliance has an interesting and unique background. Bell Atlantic and Virginia Tech had worked together in a partnership arrangement in the mid 1980s, when they jointly tested and deployed new network technology. That project was viewed as successful by both players, pointing the way to future potential joint efforts.

However, in 1986–1988, a setback in the relationship between Bell Atlantic and Virginia Tech occurred. During that time frame, Virginia Tech implemented a major upgrade of its on-campus communications systems, which resulted in the replacement of several thousand lines of telephone and data service by the installation of the large-scale IBM/Rolm digital phone switch. That upgrade caused some consternation on the part of the telephone company, from a business perspective.

In 1988, university representatives approached executives in several computer, telecommunications, and cable TV companies to ascertain their interest in a partnership effort with Virginia Tech to create a community network project. Initially, the primary players in such a network, the local telephone company and the local cable TV company, showed little interest.

In 1989 and 1990, university officials initiated discussions regarding the concept of a community network project with members of the Town of Blacksburg and its communications committee, which at that time was focusing on the local cable TV franchise agreement. The option suggested for consideration for creating such a project was an authority modeled after the Virginia Tech, Blacksburg, and the City of Radford Sewer and Water Authority. Although the authority model generated little enthusiasm, the concept of developing a community-wide information network elicited a significant amount of interest.

It is interesting to note the initial goals proposed for such a project, which were the following:

- To create a community testing ground for the 21st century learning environment;

- To create replicable community models for low-priced access to advanced, high-bandwidth communication and for universal network and computer literacy.

A concluding statement in the original notes from the presentation of those goals stated the following: "The true measure of success in this project will not be the number of consumers of information services and products, but will be the number of community producers in the proposed environment."

Shortly after the last public discussion of the community information network concept, Bell Atlantic executives approached university management in the fall of 1990 about a possible collaboration. The Bell Atlantic group proposed the possibility that the university and the telephone company, working together, could provide state-of-the-art communications capabilities to off-campus students similar to the capabilities that Virginia Tech was providing to its on-campus students.

The proposal was of great interest to the information systems management at the university, and talks continued for more than a year as various options were discussed. In the early stages of those discussions, it was clear that the Town of Blacksburg had a stake in the project, and involvement from town officials was solicited and achieved.

In January 1992, a press conference was called, and the chief executive officers of the three partners described a vision of the future. Hugh Stallard, president of Bell Atlantic of Virginia; James McComas, president of Virginia Polytechnic Institute and State University; and Roger Hedgepeth, mayor of the Town of Blacksburg, spoke about an "electronic village" to a room crowded with reporters. They were joined by other dignitaries, including Congressman Rick Boucher, from Virginia's Ninth District. The press conference concluded with the affirmation that the partners would study the concept of an electronic village and determine its feasibility.

That press conference lent credibility to the project, and the partners spent the next 12 months studying options for moving ahead and answering some of the following basic questions:

- What technologies might the telephone company employ?
- What services could the town and university provide?
- What kind of software might enable access to those services?
- What kind of organizational structure would the village have?

Many difficult issues were raised and addressed during those 12 months. Primary among them was the need for a more specific definition of the role of each partner and the concomitant required investment. Many tough discussions, or as statesmen might phrase it, "open and honest exchanges of ideas," ensued. In the end, however, Bell Atlantic had the courage to move ahead and the foresight to view the project as an experiment in the provision of potential future services.

In January of 1993, another press conference was held, with the same presidents and dignitaries presiding. This time, however, they announced that the project would move ahead, with an initial deployment occurring in the fall of 1993.

The three members of the alliance each brought unique values to the electronic village, and each sought to benefit from the project in different ways. Bell Atlantic, for example, brought an existing and evolving telecommunications infrastructure to the project. Copper and fiber optic cable, a Lucent Technologies (formerly AT&T) 5ESS digital switch, and an experienced staff of field engineers and support personnel contributed significantly to the success of the electronic village. In addition to these typical (for a telephone company) resources, Bell Atlantic of Virginia has participated in not so typical ways. For example, significant expenditures were made in Ethernet hardware, network management software, routers, and servers. By the beginning of the operational phase in October 1993, 500 Ethernet connections at four apartment complexes in town were activated. By the summer of 1995, an additional 100 ports in two more apartment complexes had been wired.

In return for its investment, Bell Atlantic hoped to gain a better understanding of future business opportunities. It is now clear that the

Internet and the National Information Infrastructure (NII) are major markets, and Bell Atlantic's participation in the electronic village can only enhance those opportunities.

In addition, the replication factor is significant. If enough can be learned from what has taken place in Blacksburg, that knowledge can be used to facilitate the development of electronic villages regionally and nationwide. In 1999, the BEV is one of about 400 active community network projects in the United States, with many more projects getting underway. As one of the first to rely entirely on the Internet as the base technology, the BEV and Virginia Tech have significant experience in developing the community-wide network infrastructure needed to support community social and economic development.

The Town of Blacksburg brought much to the electronic village project as well. The town serves as a focal point for local involvement, bringing together broad public interests: schools, libraries, health care providers, and recreation facilities. There are more than 350 local businesses in Blacksburg, including: banks, grocery stores, travel agencies, and book stores. The town enables a broad range of participation and helps to keep the interest and excitement level high. It is that broad participation, engaging every part of the community, that sets the BEV project apart from many of the technology trials that came before it.

The town also brings a tremendously important resource: town information. The town provides its citizens with electronic access to a wide array of information, including bus schedules, council minutes, street closings, and fee schedules. Residents can file vacation reports with the Blacksburg Police Department electronically, and town officials are available via e-mail and experimental monthly online chats.

The town has gained much from the project as well. There is the obvious potential from improvements in several areas, including quality of life and participatory government. Today, citizens routinely correspond with town officials via e-mail, and use the town Web site to obtain information about town activities. More significant, however, is the economic development made possible by the project. By the end of 1998, more than 24 new businesses providing Internet-related services had located in Blacksburg.

As the third partner, Virginia Tech has brought to the project a history of experience with the Internet and network-based applications, a wealth of university-related information already in electronic form, and a constituency of 30,000 faculty, staff, and students, most of whom are participants in the village.

There are great opportunities in the electronic village for Virginia Tech. First, as the BEV has grown and become a part of everyday life in Blacksburg, the network connectivity afforded by the project has been incorporated into undergraduate- and graduate-level instruction. Students communicate via e-mail with each other and with professors, and homework is now routinely assigned and turned in electronically. In the fall of 1998, all incoming freshmen were required to start Tech with a computer, completing an effort that began in 1984 when all engineering students were required to purchase computers.

Second, the electronic village truly enables research opportunities at the university. Researchers in the education, engineering, computer science, communications, psychology, sociology, and numerous other departments have undertaken research projects. Research questions include those that address methodology and a project's impact as a whole:

- How will the community be affected?

- What are the socioeconomic issues?

- How is learning affected?

- How can we evaluate and measure those changes?

Third, the electronic village is a forum for the university outreach program. In many ways, this project can be viewed as a prototype for outreach programs of the future. In local government, K–12 education, continuing and adult education, health care, small businesses, and nearly every other facet of outreach, this project provides a means to test new and innovative approaches.

Unique features of the BEV

Other important features of the BEV include:

- Community-service orientation;

- Whole-community or "open-tent" approach;

- It is a network-based as opposed to a host-based system;

- A fundamental goal of achieving critical mass.

First and perhaps most important is the issue of services orientation. From the very beginning, the electronic village was envisioned as a mechanism to provide the broadest range of services to the most people in the community. It has never been seen as primarily a technology trial, like some fiber-to-the-home or gigabit testbeds. While some new technologies have been tested in the BEV, the emphasis has always been on the provision of useful information to the end user.

Second, the BEV involves participation by all sectors of the community, including university, K–12 education, government, private citizens, health care providers, and local businesses. This broad participation presents some formidable challenges in terms of coordination and technical support, but it also offers some unusual potential. By taking the open-tent approach, the data network can be perceived as a utility, as opposed to being associated with one or more specific applications. In addition, the willingness to accept and encourage business participation is truly unique and presents a smorgasbord of unique opportunities.

Third, the BEV is network based, as opposed to host-based. In other words, the BEV is not an enhanced bulletin-board system with Internet access. In fact, the electronic village is a microcosm of the Internet, with thousands of computers in Blacksburg utilizing the Internet locally and globally to communicate with friends, family, and people around the world.

Internet-based services like e-mail, the World Wide Web (WWW or just Web), forums, and discussion groups are provided by the BEV project group at the university. All of these services can be distributed, and additional services can be made available by literally anyone in town, merely by connecting a server to the network. Hundreds of WWW servers are located in Blacksburg; most of them are run by students and other residents of the town who simply want to experiment with this new publishing phenomenon. Thousands of town residents have their own home pages on the servers.

Fourth, and perhaps most important to the success of the project, is the goal of achieving a critical mass of network users. The term critical mass is somewhat vague, but in this context it is that point at which there are so many users that the existing applications become truly useful and another whole array of applications becomes feasible. By the fall of 1997, studies indicated that more than 80% of Blacksburg residents were using the Internet routinely, making Blacksburg the first in the world to achieve such high per-capita use.

For example, over the course of the past 17 years, most corporations have adopted e-mail. In a company of 1,000 employees, e-mail initially may be used by only 10 systems programmers; it is fairly useful in that context for the dissemination of technical information. In the next stage, e-mail begins to be used by computer-knowledgeable employees and "experimenters," numbering 100 or so. In this stage, the proof of concept starts to show through, and the potential value of e-mail begins to make itself known throughout the company.

Finally, e-mail permeates the company, with well over half the employees using it on a daily basis. When that happens, e-mail becomes the preferred method of both interpersonal and corporate communications because it is easier and quicker than conventional memos and reports. In addition to normal memoranda, e-mail begins to be used for other purposes, such as purchase orders and work orders, and personnel actions could all be based on the e-mail infrastructure.

In the same way, a critical mass is achieved in the electronic village, and a whole set of new services are enabled. Put another way, so many folks are utilizing data connections that people assume the availability of a network connection as a given. Teachers and doctors utilize the Net in their everyday interactions with parents, students, and patients. Local businesses naturally want to use the Net to ply their wares.

In Blacksburg, civic groups have now begun to abandon paper distribution of newsletters and related information; with most (if not all) their members online, scarce funds can now be allocated almost entirely to the primary civic purpose of the group. Formerly, the single biggest expense for many civic groups has been the group newsletter (printing and mailing costs). Businesses in Blacksburg now frequently run a mailing list (listserv) for customers, offering specials and sending out notices when new products arrive.

New services have emerged, many of which have no analog in the current environment of service provision. For example, the electronic yellow pages now online in many versions is completely different from existing printed yellow pages. It is easy to provide multiple indexes to businesses, and individual businesses are not limited to a fixed space for information: Entire product catalogs and online order forms become part of the yellow pages listing.

Early growth of the BEV

By the spring of 1995, more than 40% of town residents were using the Internet, and more than 60% were using e-mail. Some residents with e-mail are obtaining that service from commercial online providers like America Online, CompuServe, and Prodigy, while some faculty and staff still use the university mainframe for e-mail services.

Over the summer of 1993, following a beta test phase and in anticipation of the operational phase beginning in the fall of 1993, much transpired. Bell Atlantic installed Ethernet connections in four apartment complexes and its central office and provided a full 10-Mbps connection to the Virginia Tech campus. Integrated services digital network (ISDN) was deployed on a pilot basis.

Virginia Tech revised the software used to connect a user's computer to the Internet, based on user feedback obtained in surveys and focus groups. Virginia Tech also provided authentication and authorization services. The university deployed a high-speed modem pool and provided Ethernet services to on-campus students. The Town of Blacksburg organized its information and made public access systems available in the Blacksburg branch of the regional library. Finally, BEV, Inc., was formed to provide community representation to the project. The board of directors of that nonprofit corporation meet several times a year to provide input to the members of the alliance and to assist in grant development.

On October 25, 1993, the BEV opened its doors in a temporary office in the Corporate Research Park, where the university staff managing the project had their offices. Demand was steady for the first year. Between 75 and 100 people per month signed up, which was somewhat less than expected, but it kept the two part-time people working in the office busy with questions about access, developing new administrative procedures, and refining software and services.

In the late spring of 1994, the first Internet business in town, a WWW server company, opened its doors. Biznet Technologies provided a place on the Web for local businesses, and after a series of business seminars conducted by the BEV group at the university, businesses began to take an active role in the village.

While the BEV project office was catering to townspeople, the university was also promoting the use of the network on campus. Faculty, staff, and students could sign up for service at the BEV office (and many did), but many also obtained access through normal university procedures. Most students (14,000 of them) live off campus, so even if they obtained their BEV software from the university rather than the BEV office, they still shared identical network services and connectivity.

In the media, 1995 was the year of the Internet, and the nationwide publicity about the new communications medium helped in Blacksburg. A watershed event occurred in January of that year, when the BEV office moved from its temporary quarters in the Research Park to Main Street, one block from the center of town. There was an immediate and permanent increase of more than 100% in people registering for service. It was clear that a physical sense of place was needed to convey the existence of Blacksburg's sense of place in cyberspace.

As demand for services rose, the lack of revenue to support those services became more critical. In the early part of 1995, the three members of the alliance and the BEV, Inc., board of directors held a series of wide-ranging discussions. It was becoming obvious that at least part of the experiment was coming to a close. Demand by the university members and BEV users in the community for dialup access to the Internet was causing severe congestion on the university modem pool. That service was priced on a cost-recovery basis designed primarily to serve the university, and now, after two years, private-sector investment in modem pool services was needed in the community to satisfy demand.

In May of 1995, the project announced that it hoped to migrate BEV users not affiliated with the university to a private modem pool by the end of the year. An energetic online discussion of this change in policy ensued, with some BEV users upset over the announced change. More important, at least six local businesspeople saw an opportunity to get into the modem pool business that the university apparently was getting out of, and by late fall two of those new ventures opened for business.

It was none too soon, because the BEV office, open only nine hours a week and staffed by just two part-time workers, was near collapse under the load. In August of 1995, nearly 400 people visited the BEV office to sign up for or modify their network service. Combined with other users coming in for technical support or just to ask questions, during the 36 hours the office was open in August, more than 16 people per hour were passing through, an average of one person every four minutes.

During the spring and summer, public relations became nearly a full-time job. Television camera crews became a common sight in Blacksburg as both national and international news coverage of the town and the project raised Blacksburg's visibility as "the most wired town in America."

In May of 1995, the Town of Blacksburg announced what was expected to be a modest economic development effort by donating $15,000 to BEV, Inc., for redistribution to Blacksburg businesses starting Web pages. At the time of the announcement, there were about 300 registered businesses in Blacksburg and about 125 listings on the BEV Village Mall Web pages (http://www.bev.net/mall/). The announcement created a torrent of activity.

Several new Web service companies started up or added Web services to existing services, and over 70 grant applications were submitted to BEV, Inc., for the grants, which could only be used to design the pages (not for monthly service fees) and were limited to $500. Thirty-five grants were awarded, but by year's end, there were more than 200 businesses with listings on the BEV Village Mall. At the start of 1996, Blacksburg businesses not advertising on the Internet were in the minority.

Ron Secrist, the town manager at the time, remarked that it may have been the cheapest $15,000 the town had ever spent. One of the requirements that businesses had to meet to receive a grant was to register (or already be registered) with the town for the purpose of paying business taxes. Ron suspected that the town would recoup most of the money through increased business tax revenues because there was a flurry of new business registrations after the grant program was announced.

Although the staff of the project had hoped for a quiet fall, 1995 was to finish with a bang. In October, the university group won a National Telecommunications Infrastructure Administration (NTIA) grant from the Department of Commerce with the assistance of BEV, Inc., and the

Montgomery County public school system. The 18-month grant provided for increased support at the local public library; a new computer lab in Auburn High School, which is in a very rural part of the county; and assistance for the City of Radford and other localities in southwest Virginia that wanted to start electronic village projects of their own.

Congressman Boucher's home town of Abingdon, Virginia (http://www.eva.org/), became the second locality in the area to have a community network project. After several months of work and numerous planning sessions with BEV staff from the university, Abingdon's home page came online in February of 1996. The City of Radford followed shortly afterward in April 1996. Tiny Craig County (pop. 4,300) began work on a community network project with the help of the BEV staff and went online in 1997 (http://www.co.craig.va.us).

On March 1, 1996, the university announced that on March 15 it would no longer sign up users for dialup access if they were not affiliated with the university. The two private modem providers had been in operation for nearly three months, and it was time to let the private sector take over that portion of the project. A major goal of the BEV—to create enough demand for network services in the community that the private sector would offer Internet access as a for-profit operation—had been achieved. By April 1996, over 500 dialup access lines were available in the area when the university service was combined with the two private providers, yet there was still heavy congestion in the evenings.

Sometime during the fall of 1995, critical mass had been achieved in Blacksburg. It is still not clear what the magic number or measurement is, but the BEV office staff observed that late in the year an increasing number of people coming to the office were making one of the following comments:

- They had tried getting connected early in the project, had had problems, and had given up, but now so many friends and neighbors were online that they were determined to see it through this time.

- They had never had much interest in computers or the network, but so many people they knew were on that they felt they must be missing something.

In April 1996, the town hall sprouted new signs on the side of the building that included not only the street address of the physical building but also the uniform resource locator (URL) for the address of the town in cyberspace (http://www.bev.net/). It was truly a sign of the times and most appropriate.

Later growth of the BEV

The change in modem providers provoked angry feelings from some town residents. What was surprising was that the complaints were generally not about the nearly doubling of access costs, but about the perceived loss of identity with the BEV. Many people changed their e-mail address to the one provided by their new Internet service provider (ISP) and were upset that they were no longer part of the "bev.net." It was clear that the project had created a new sense of awareness and belonging in the community.

As part of Virginia Tech's commitment to networking and learning at a distance, the university brokered a far-reaching partnership during this time to implement a statewide broadband network. Net.Work.Virginia, as it is called, is a broadband asynchronous traffic mode (ATM) network that carries Internet data, high-quality video, and voice telephony traffic simultaneously. Net.Work.Virginia (http://www.networkvirginia.net/) provides flat-rate service anywhere in the state, regardless of distance or location. The BEV group provides some of the backbone network management services (DNS) for the system, and in late 1998 there were about 350 educational institutions, public agencies, and municipalities connected statewide. Net.Work.Virginia has become a model for regional networks, and is one of two network architectures under consideration for next generation Internet designs.

In late 1996 and early 1997, the BEV group began noticing network traffic problems emanating from apartment Ethernets. Virginia Tech students were making such heavy use of campus servers that severe congestion was creating noticeable degradation in service. In analyzing the problem, the BEV group realized that in a distributed community-wide network with many access providers, a local routing and switch point was going to be required to keep local traffic local. The apartment Ethernet problem was caused by data packets that had to travel over large parts of the East Coast of the United States just to reach the other side of town. This happened because the interconnect points for backbone Internet access providers were located only in major cities, not in small towns and cities.

Blacksburg now had enough local traffic to need its own local routing point, commonly called a network access point (NAP). In the past, NAPs were commonly set up and managed by ISPs, but the BEV wanted to create new business and service opportunities as well, and so planning for a community NAP began in earnest. One of the first changes that took place was in the name—NAPs normally provide only Internet-based routing, but the Virginia Tech and BEV vision was to provide a switching and routing point to carry all kinds of broadband services, including voice telephony and video, so the name was changed to multimedia services access point (MSAP).

The MSAP became part of Virginia Tech's plans for new, regional network architectures, and ATM switching capabilities were added to the overall MSAP design. The MSAP became operational in Blacksburg in early 1999, with several local ISPs as the first customers. Eventually, most communities will have similar facilities, and it is expected that the MSAP, like other BEV access services, will create new businesses and jobs in the area because of the reduced cost of access to local customers.

In 1997, the BEV's aging Web interface was overhauled completely. Web traffic statistics from the previous three years were carefully analyzed, and a new page design was implemented that reflected what people most commonly used. A major portion of the BEV home page was devoted to timely community news and announcements.

In 1997, the BEV group also began the transition to a cost recovery basis for some of its services. Until then, many of the BEV services provided to the community had been free, but this model was clearly not sustainable. By early 1999, it had become clear that a broader base of support is needed to create sustainable community-wide networks. It is not possible or reasonable to charge directly for important services like the community Web site. The BEV group is still exploring various economic models, but a broad, community-wide support mechanism seems to be the correct strategy.

By early 1998, there was increased development of residential Ethernet services in Blacksburg, and Bell Atlantic and the BEV group made the decision to move the apartment Ethernet experiment into the private sector. Since 1993, Virginia Tech and the BEV group had been managing the Bell Atlantic infrastructure in six apartment complexes.

It was expected that local ISPs would take over that business, but once again, surprises were in store. Several apartment owners collaborated to form a new company dedicated to providing Ethernet services, and this new company took over most of the apartments with little fanfare. Local ISPs did pick up the rest of the business. As of July, 1998, the BEV was finally out of providing Internet access, a major milestone in the project.

The BEV has been criticized for providing access in the first place; critics have noted that that is a job properly left to the private sector. However, it is important to note two things:

■ The BEV offered Internet access by modem and by Ethernet at a time when the private sector was offering no equivalent service, and at a time when the private sector was unsure such services were economically viable.

■ When it was clear that such services were economically viable, the project divested the services and in the process created several new entrepreneurial startup companies and created dozens of new, good-paying jobs in the area.

In 1997, as use of the BEV Web site and related services continued to increase while the BEV staff shrank slightly, the group embarked on an ambitious project to move most of the Web site to a database-to-Web publishing system. Doing so would decrease the amount of staff time needed to maintain community information while giving users in the town more control over when and how information was posted. Work proceeded slowly, but by early 1999, the Village Mall was converted from a series of "flat" Web pages maintained by hand to a fully automated publishing system with a built in search engine, better indexing of businesses and services, and improved listing information.

Other services that will be implemented in 1999 include a community directory to make it easy to locate e-mail and Web addresses for other community members (a "white pages" system), dramatically improved free Web sites for personal use, and much improved services to community and civic groups.

The future of the BEV

The BEV project will never end, because the electronic village belongs to the people of the town, not to the university, the local government, or the ISPs. The value of a community network is the new channel of communication it offers to a community. It is an opportunity for members of the community to communicate with each other about issues and problems of common interest. In Blacksburg, everything has changed, and nothing has changed. The town looks much the same as it did six years ago, and the people of the community still get together in the coffee shops, restaurants, and bars, just as they did three years ago. As an interesting side note, there are now more stores and restaurants in Blacksburg than at the start of the project. Although we cannot say this is due to the BEV, it seems clear that in Blacksburg, a community network and the availability of online shopping didn't affect local retail stores in a negative way.

It is interesting to evaluate the BEV in the context of the original statement of goals in 1989.

- The majority of the citizens of Blacksburg can now be regarded as computer- and network-literate.

- Many communities in the state, the nation, and the developed world have interacted with key people involved in this project with the idea of replicating the Blacksburg model for an electronic village in their own community.

- Thousands of citizens in the electronic village have become producers of information services and products serving not only the local community but also their extended world community.

Some people fear the network and the changes that it will bring, while others regard it as a panacea for the ills of the nation. Technology is only as good as the people that use it. In Blacksburg, the network has provided a new way for people to talk about themselves and about their community.

For the future, Blacksburg will continue to experiment with new and innovative services. With most citizens online, it is now possible to think about a "new democracy," in which fast, convenient access to information about important civic issues and new kinds of community online community discussions of those issues leads to better community decision-making as the residents of Blacksburg grapple with complex 21st century problems of growth, development, education, and quality of life.

3

The Architecture of a Community Network

by Andrew Michael Cohill

Access and services

A discussion of any kind of network, including the Internet, can be reduced to four things:

- People use networks to communicate with other people. Networks enable human-to-human communications. This human communication is the primary reason we design and build networks. Community networks are developed to enhance the ability of people in the community to share ideas, discuss issues, and solve problems.

- Services describe what you can do once you have established access. Some of the services that people use in Blacksburg include e-mail and the World Wide Web. You cannot use any Internet service until your computer has established access to the network (usually by running a small piece of software installed on your computer).It is easy to confuse the two ideas; having network access does not necessarily mean that you can use any of the services, and e-mail software installed on your computer will not work without access.

■ Infrastructure is how you connect to the physical and logical infrastructure of wires, cables, network equipment, and computers that comprise the network, wherever they may be. For example, many people in Blacksburg access the network with a modem and a plain old telephone line. Other Blacksburg residents access the network using direct connections that are completely separate from normal telephone lines.

■ Content is developed by people. Content takes many forms, based on the kinds of services available on the network. It includes e-mail, the World Wide Web, discussion forums, and mailing lists as some of the most common content delivery systems. Content is stored on computers that deliver it in a specific format (the service) using the infrastructure of the network to deliver it to people.

This chapter tries to describe the various parts of a typical community network without going into too much detail. Later in the book, Luke Ward's chapter on network technology provides a comprehensive description of many of the terms and ideas introduced in this section.

A typical community network

The community network depicted in Figure 3.1 is just one of many possible network configurations. Figure 3.1 contains one of the most common access methods in use in Blacksburg, but it has been simplified in some ways to keep the diagram easier to explain and to understand. For more detail on the equipment and concepts discussed here, see Chapter 10.

Item A: computer on the Internet via modem

Most people, wherever they live, access the Internet using an ordinary telephone line[1] and a modem. Modems take the digital information from

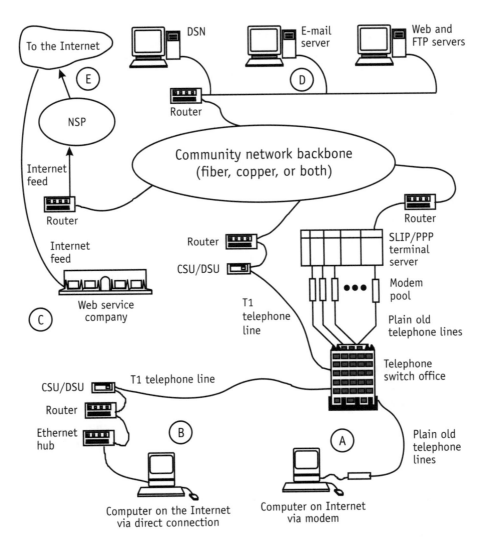

Figure 3.1   Architechture of a community network.

the computer, convert it to a signal transmitted over a phone line, and convert it back to digital information at the other end of the phone line.

One modem is connected to your computer in your home, and the modem on the other end is part of a group of modems (usually called a

---

1. Often referred to as plain old telephone service (POTS).

modem pool). Modem pool providers lease telephone lines from the telephone company (one for each modem in the modem pool), so that many people can dial the modem pool at the same time. Each modem in the modem pool takes the information being transmitted from your computer and passes it to a piece of electronic equipment called an access server.

The access server takes the many separate information signals from the modem pool and converts them to packets of information in Internet format—often called Internet protocol (IP). Each packet has address information attached to it, and once on a leg of the Internet network, it is passed through various routing points until it reaches its destination.

The process works in reverse as well. When a computer attached to some part of the Internet sends information to your computer (perhaps a piece of e-mail or a Web page), a series of packets with the address of your machine (your IP address) is sent across the network until it reaches the access server that knows you are connected via modem. The information is converted to modem format, passes down the telephone line, is converted back to digital information by the modem attached to your computer, and is then converted into the information that you see on your computer screen.

There are now alternatives to dialup modem access in many communities. ISDN (Integrated Services Digital Network) is similar to modem access (it uses phone lines and similar electronics) but operates about twice as fast as the now common 56-Kbps modems. ISDN is relatively expensive and is being replaced by digital subscriber line (xDSL) services. xDSL also uses your phone line and modem-like electronics but can provide speeds 5–10 times faster than modems. Prices for xDSL average about $60/month nationwide and represent a very good value. The cost can be much higher and is related to the amount of upstream bandwidth available. xDSL will eventually replace modem access for many people. Both ISDN and xDSL are distance-sensitive technologies, which means you must live within a few miles (2–5, typically) of your telephone switching office. This is often a problem for rural communities.

Item B: computer on the Internet via direct connection

In Blacksburg, it is also quite common for users to have direct connections to the Internet at home or at work. No modem or telephone line is required, but users have a separate jack in the wall (similar to a phone jack, but with different wires) that is connected to an Ethernet[2] network. In Blacksburg, the most common kind of Ethernet is 10BaseT, which is inexpensive copper cable. Computers hooked to an Ethernet network must have either built-in Ethernet connectors (as with most Macintosh computers) or an added Ethernet card (as with most Windows computers). The cards usually cost less than $75.

The computer in Figure 3.1 is connected to a port in an Ethernet hub, which concentrates the packets of information from many computers (a hub may have as few as four and often has 64 or more ports). All of the packets are retransmitted on a single Ethernet line to a router (which routes information packets to and from their destination).

In Figure 3.1, all of the information from that local Ethernet network is then routed through a high-speed phone line (which could be copper or fiber) called a T1 line. T1 lines are able to transmit digital information at 1.5 million bps. Note that a CSU/DSU is used to convert the digital information to T1 format. The T1 line travels to the telephone company switching office, just like a regular phone line, and a second T1 travels out from the switch office to another CSU/DSU and router that is connected to the community network backbone (which carries information in IP format). Numerous apartment buildings and offices in Blacksburg are hooked up to the Internet in this way.

Nationwide, cable modem technology is being introduced in many communities. Cable modems take advantage of Ethernet by carrying the signals over your existing cable television system (CATV). There are two forms of CATV systems: one-way and two-way. One-way systems represent more than 80% of the installed cable systems in the country, and using cable modems with one-way systems means that you must also use a telephone line and standard modem. The very fast Ethernet connection via the CATV systems sends information to your computer several times faster than telephone modems (speeding downloads of graphics, Web pages, and software), and the slower telephone modem is used to send

---

2. Ethernet is an electronic standard for transmitting digital information. All it really means is that there is a second set of cabling in the wall in addition to the standard phone cable. Some kinds of Ethernet cable look just like phone cable and do not cost much more.

requests for information out from your computer. Two-way cable systems allow transmission in both directions, so no telephone modem is needed. Few communities currently have two-way systems, and in rural areas, some companies may be slow to replace existing one-way systems because of the cost.

Item C: community servers

In Blacksburg, hundreds of information servers are located all over town. Servers are well named; they serve (send) information to people whenever they receive a request. When you check your e-mail, your personal computer sends a request to a mail server located somewhere on the network (e.g., America Online, CompuServe, or your local Internet provider); after receiving the request, the e-mail server sends (serves) your mail to you. Each time you click on a link on a Web page, it creates a request sent to a World Wide Web server for the information associated with that link. The word "server" is used interchangeably to mean both the computer (hardware) and the computer program (software) that actually executes the instructions.

Anyone with a computer and access to the Internet can run a server. Most servers are connected to the Internet with full-time direct connections. It is possible to run a server with a modem connection, but other people would only be able to access the information on the server when the computer is linked via the phone to the Internet. Because modems are slow, the amount of information the server could deliver would be limited by the speed of the modem.

Figure 3.1 shows three of the most common kinds of servers used in community networks.

- A domain name server (DNS) is required in any local or regional Internet segment as an essential service. DNSs translate Internet names such as mail.bev.net into numeric Internet addresses such as 128.173.4.30. Every Internet application (i.e., e-mail and Web browsers) on a computer on the Internet, including all personal computers, constantly use a DNS to ensure that all information packets are addressed correctly. DNSs are usually set up and managed by the local Internet provider, but this could also be a service of a community network. In Blacksburg, the BEV provides name services for all public and private organizations in town except Virginia Tech (which runs its own name server).

■ An e-mail server is also an essential service. E-mail servers function like a post office with post office boxes for each user. Mail sent to joe.smith@bev.net is stored on the BEV mail server named mail.bev.net until Joe (using the e-mail software on his computer) requests his mail from mail.bev.net. The mail server responds by sending Joe's mail to him; it is copied onto the hard drive on Joe's computer and is usually deleted from the server. In Blacksburg, Virginia Tech provides e-mail for both the community and the university using a single machine. On busy days, more than 250,000 e-mail messages may be processed and delivered by the machine (in a town of 36,000 people). Mail servers and e-mail addresses are usually provided by your local Internet service provider (ISP).

■ World Wide Web and file transfer protocol (FTP) is a way of moving computer files between two computers on the Internet. Servers are core services for a community network. Web servers deliver information that can be viewed with full-color graphics and text (just like paper documents), audio, video, and other multimedia services. In Blacksburg, the BEV and Virginia Tech run the primary Web servers used by the community, but hundreds of residents and businesses also run Web servers. FTP servers are commonly used to allow users to download software and large files. The BEV runs an FTP server to make it easy for BEV users to download updated copies of BEV software without making a trip to the BEV office with a diskette.

Item D: national service provider

The national service provider (NSP) plays an important role; it connects the network and computers in a community or region to the rest of the Internet. The local ISP (community network or private company) buys an Internet feed from an NSP like MCI, Sprint, or AT&T. Information flows up and down this link (a special high-speed phone line that may be copper or fiber) as requested.

Item E: Web service company

In Blacksburg, dozens of private companies are in the business of leasing space on their World Wide Web servers to businesses and other organizations that do not wish to run their own servers. Some of these companies have their own private feed to the Internet. Anyone in the community can access the information on these servers even if the company has an Internet feed separate from the community network because the system of domain name servers (DNSs) and routers uses IP addresses to ensure that the information packets are delivered properly. Geography is irrelevant on the Internet!

Summary of network services

Table 3.1 provides a list of network services commonly used on the Internet that are required or likely to be of interest to community networks.

Network support

As described previously in this chapter, DNSs perform an essential Internet service. Name servers translate English language Internet names (mail.bev.net) into numeric Internet addresses (128.173.4.30).

User identification

Authentication is used by many Internet services (e.g., e-mail) to establish that you are who you say you are. This is usually done with a userid (jsmith) and password (usually 5–8 characters). The userid is public, and the password is private. In Blacksburg, the university and the BEV use an authentication server that allows multiple Internet services (i.e., e-mail, Web, and news) to authenticate users with a single userid/password combination. This is convenient because users normally have to remember only one userid/password combination; in many places, users may have several userid/password combinations to remember.

Authorization is used to determine what you are allowed to do once your identity has been established (authentication). In Blacksburg, the authentication server also provides authorization services.

Table 3.1
Summary of Network Services

Service	Essential	Core	Enhanced
Network support			
Domain name service (DNS)	×		
User identification			
Authentication	×		
Authorization	×		
E-mail services			
E-mail	×		
Listservs		×	
Forwarding	×		
Directory services		×	
Information distribution			
Web server (organizations		×	
Web server (personal pages)			×
Secure server			×
FTP/fileboxes		×	
News			
News (full-feed)			×
News (limited-feed)		×	
News (local only)		×	
News (private groups)			×
Online conferences and chat			
Online conference system (Web-based)		×	
Chat rooms (Web-based)			×
Chat rooms (IRC-Java-based)			×
Advanced technology			
Internet radio			×
Videoconferencing			×
Telephone services		×	
Low power radio		×	
Video streaming			×

E-mail services

E-mail is an essential service. It is the most common Internet service used in the Blacksburg community by a wide margin. Much like paper-based mail, there is a message body and an address. Sending a piece of mail is as simple as typing a few sentences, adding a brief address (jsmith@bev.net), and clicking a button. E-mail is normally delivered to recipients in a few seconds or, at most, a few hours (no matter where they live in the world).

Listservs are also known as mailing lists. A listserv takes a single piece of mail sent to an e-mail address (e.g., cats-L@listserv.mail.com) and resends it to a list of e-mail addresses. In Blacksburg, the BEV Seniors group uses a mailing list to keep in touch with its 100 members.

Forwarding works just like forwarding at the post office. If your mail server provides this service, you can set forwarding to send your e-mail to another e-mail address. This is convenient if you have more than one e-mail address or if you change your e-mail address. Forwarding can be a valuable community service, because it can be used as a way to give everyone in the community a single e-mail address (e.g., bev.net) even if their e-mail box is located on another server.

Directory service can be provided in a variety of ways, but it is useful if some group in the community provides a place where users can look up the e-mail address of other people in the community. The telephone company provides this service for telephone numbers, but in most communities people have e-mail addresses from many different companies (e.g., jsmith@bev.net, jdoe@aol.com, and susan@prodigy.com). A community directory can help locate friends and neighbors in cyberspace. Directories can also list the addresses of personal home pages, private FTP sites, and other information such as telephone numbers and street addresses. Any directory service should give people the option of leaving out certain information like street address or phone number.

Information services

Web servers, described previously in this chapter, transmit full color text and graphics to any computer on the Internet with a World Wide Web browser. Web browsers are also have sound and video capability. It is now common to use databases to publish content via the World Wide

Web, and many sites, including the BEV Web site, make extensive use of databases to provide search capabilities and to simplify content management.

Secure Web servers use encryption to scramble information while it is in transit across the Internet. Many Web browsers support encryption, and it is most commonly used when transferring financial information such as credit card numbers, corporate data, and banking data. Secure servers are an important part of ecommerce initiatives because the encryption is used to hide personal information like credit card numbers.

FTP servers and fileboxes are used in Blacksburg to give people access to software, large data files, and other information. Most computers that have a Web server on them also support FTP, so that users are able to put new information (files) on the Web server. That is, FTP is used to send the information files to the computer, where the Web server makes the data available to others with Web browsers.

News

Usenet, a group discussion service, is one of the oldest public services on public networks. Usenet existed long before the popularity of the Internet. Usenet is text-based; Usenet messages are composed just like e-mail, but instead of being sent to a single e-mail address, messages are sent to a news group at a local news server. This news server, in turn, makes the messages available to anyone with a news reader. There are more than 20,000 Usenet groups worldwide; most are in English, but there are country-specific groups in French, German, Swedish, Norwegian, Japanese, and many other languages. Many newsgroups are computer- and network-related, but no matter what your hobby, occupation, passion, or problem, there is probably a Usenet group where you can meet people with similar interests.

Local Usenet groups are a place to talk about local activities. There are about a dozen Blacksburg groups (e.g., bburg.general, bburg.bev. announce, and bburg.auto.repairs) where the community discusses a wide variety of topics and issues. Local newsgroups are an important part of community networks, because they provide a local public gathering place for discussion.

Private Usenet groups have restricted access (only select people are allowed to read and/or post messages) and can be used to support classroom activities, businesses, and community groups.

Online conference and chat service

Internet relay chat (IRC) is similar to Usenet in that it is also a public message system where many people can communicate. It is different from Usenet in that chats take place in real time, like a telephone conference call. A chat room (also called a forum) is a group of people meeting in cyberspace at the same time. People communicate by typing messages on their own computers, which are immediately posted to the chat room where they are read by everyone else in the room. There is a wide variety of chat services, including some that use graphics and photographs to make them more interesting though people still communicate via text.

Web-chat servers allow use of a Web browser to participate in chat sessions without special chat software. Because Internet users are familiar with Web browsers, it is easier to support Web chats. In Blacksburg, several Web-based chat rooms are available for community use.

Chat is primarily used for nonbusiness, social activities and is not well-suited for online discussion of community and civic topics. Online conference systems are much more sophisticated than chat and have a broad array of features that make them ideal for "town hall"-style online discussions. The BEV introduced online conferencing to Blacksburg in 1999.

Advanced technology

Internet radio is a new service that delivers taped or live audio information (just like radio) over the Internet instead of via radio waves and antennas. This service works with modems, but sounds a bit like AM radio with respect to quality. In the future, there will be thousands of radio channels broadcasting music, talk shows, weather, news, and all kinds of special information via this technology. The cost of providing Internet radio has declined dramatically, and this will become a common community network service in the future.

Videoconferencing over the Internet lets two or more people talk (audio) and see each other (video). This service has been in experimental use in Blacksburg for more than two years, and in the fall of 1996 the BEV group and the Virginia Museum of Natural History began regular science lecture programs to local public schools using this system. Macintosh computers are particularly suited for this, as most models have plugs on the back for a video camera. By 1999, videoconferencing over the Internet was still only suitable for hobbyist style and casual business/school use because of bandwidth limitations. Although the computer hardware and software technology has improved dramatically, lack of adequate bandwidth continues to be a persistent obstacle. Virginia Tech began using a very high-bandwidth statewide ATM (see Chapter 10) network in 1997 to provide distance learning and high-quality videoconference services, using dedicated digital video (i.e., not the Internet) channels.

Telephone service has become common on the Internet, with several companies offering inexpensive software (under $50) that enables two-way audio conversations over the Internet. Like Internet radio, this works fairly well with a modem and has become a popular way to save money on long-distance telephone calls. Voice over IP (VoIP) uses your computer and your Internet connection to allow you to have a two-way voice conversation with someone else. In some cases, this may be with someone who is also using a computer and an Internet connection, but in many cases, your conversation is transported partly over the Internet and partly over normal telephone circuits.

The most common use today is for overseas phone calls. Someone in the United States can make an international call using VoIP; typically, the Internet will be used to carry the portion of the conversation between the United States and the destination country, and there, a VoIP service provider switches the call to a normal phone circuit. The person receiving the call uses the telephone, and the person in the United States is using their computer (with microphone and speaker).

In the future, when we all have good full-time Internet connections in our homes and businesses, we will make all of our phone calls over the Internet. Some phone companies have petitioned Congress to block use of this service until the relationship between it and regular telephone services can be determined. By 1999, use of this service was widespread. In Blacksburg, which has a large international student population, VoIP telephone software was in wide use to make long distance phone calls overseas. Because part of the call must be carried over conventional telephone circuits, this service is likely to provided only by private businesses.

The BEV software tools

The BEV provides a standard set of software tools to users. Both Macintosh and Windows computers are supported, although the Mac software is much easier to install and use due to the superior network capabilities of the Macintosh. All of the BEV software is freeware (it can be distributed freely without fees or licenses) or has been licensed for use.

Serial line IP(SLIP) or point-to-point protocol (PPP) software establishes your network connection to the Internet if you use a modem. Before you can use any of the Internet software listed below, you must start your SLIP or PPP software. Once the SLIP or PPP has established the connection, your computer is connected to the Internet, and you can access any other computer on the Internet that permits public access or has a server program running. To leave the network, you must close all of your applications, return to the SLIP or PPP software, and shut down the connection. SLIP is an older technology, and many ISPs have replaced it entirely with PPP.

Eudora Lite is an e-mail client[3] for exchanging e-mail with others. Your e-mail address combines your userid with the designator for the post office computer where the network should deliver your mail until you pick it up with Eudora.

An FTP client allows you to transfer files among computers. The many anonymous FTP sites on the network provide open access to files and materials of all sorts. With an open connection, you can change directories on the remote machine or retrieve files from it simply by clicking on the name of the file or directory.

A World Wide Web browsing tool is a piece of software that lets you use a multimedia information service that represents an expansion over the capabilities of Gopher. The Web is divided into pages. Any part of a page can be a link to another Web page, a Gopher menu, a picture, a

---

3. A client is just a piece of software that runs on your computer.

movie, a sound clip, or just about anything. Netscape and Mosaic are the two most popular Web browsers.

Telnet software connects you across the network directly to another machine so you can log into it. (You must have login access to that machine.) There are unrestricted guest telnet sites around, but no specific tools for finding them. Many are linked through Web sites, and Web browsers will launch telnet when needed. A popular use of telnet is to access "shell" accounts on Unix systems, where command line interfaces provide access to e-mail, operating system functions, and software development environments.

A Usenet news reader provides access to the Internet's hierarchy of newsgroups, public-access discussion groups (of which there are thousands), each dedicated to a particular topic. Usenet readers vary widely in quality and design.

Service models for community networks

Some representative costs for various aspects and functions of a community network are listed below. All of these figures are rough estimates based on experience. As a general rule, expect the cost of equipment to decline somewhat and expect the cost of staff to increase over time. The items discussed here address only the cost of delivering services to the community.

Level 1

This is the simplest kind of community network. The town government or a group of citizens leases space for a community WWW site. Access is provided by local or national dialup (SLIP/PPP) modem pool providers. Other services like mailing lists can also be purchased from ISPs. This approach is inexpensive and even the smallest community can afford it, as modest Web sites can cost as little as $50 per month. Each individual bears the cost of access himself or herself by paying, on average, about $25 per month for dialup access to the Internet (see Table 3.2).

There is a wide variance in costs because there is more than one way to provide the resources needed to support the network. For example, if the local library supplies space for training, that cost (to the community

Table 3.2
Level 1 Model—Lease All Equipment, Services, and Technical Management

Startup Costs		
Item	**Who**	**Cost**
Initial design of community site	Web design firm or local help	$500–$5,000
Ongoing Costs		
Item	**Who**	**Cost (Monthly)**
Community Web site	Internet service company	$25–$200
E-mail mailing lists (2–3)	Internet service company	$25–$50
Content maintenance	Part-time staff	$100–$1,000
Education/tech support for residents	Staff	$100–$2,000
Office space and utilities	Per staff member	$300–$500

network) may be very low because the cost of the space is covered by the library's budget. Similar options exist for staff as well. In Radford, Virginia, one of the local librarians manages the Web site half-time, so his salary is also covered by the library's budget. In another community, it might be that a city employee or even a civic group takes on some of the training and system administration tasks to help save money.

Level 2

Another option for small and medium-sized companies is to purchase a computer and WWW server software, lease an Internet feed for the machine, and run a community Web server. This gives the community more control over the Web site, and it may be less expensive in the long run if public-access computers are also connected to the Internet feed. Web server "appliances" are now common; these are turnkey computer systems designed to provide a complete Web hosting system at very low cost and with very little technical expertise.

With this scheme, most users still access the Internet via modems. Schools and businesses in the community may also have direct connections to the Internet, but each is purchased separately from local or national IAPs. A single computer for a Web server (preferably a Macintosh for ease of use) will cost about $1,500 to $4,000, and ongoing costs for the Internet feed will vary depending on the size and type of feed. However, you can expect to pay between $500 and $2,500 per month for a full-time dedicated connection.

Level 3

Level 3 is much like level 2, but the community invests in more computer hardware and software so that all essential services and some core services can be offered by the community network rather than by the private sector. An advantage to providing more services is that free access to e-mail accounts, Usenet news, and local discussion groups can be provided in the public library, in public schools, and other public-access locations like a community center at much lower cost.

At this level, a part- or full-time technical support person will be required to manage the two or three computers used to provide these services. Distributing e-mail accounts will also mean that an office space and at least part-time staff will be needed to manage administrative work related to e-mail accounts. An initial investment of $10,000 to $25,000 in computer and network hardware will be required. Staff and office expenses could total $15,000 to $30,000 per year. The Internet feed will vary in cost depending on the size and type of feed, but expect to pay between $500 and $2,500 per month for a full-time dedicated connection (see Table 3.3).

Level 4

At this level, the community makes a substantial capital investment of $50,000 or more to become a local ISP. In addition to the computers and network equipment purchased at level 2, a high-capacity Internet feed and additional network equipment is acquired so that sites around the community can be fed with high-speed direct connections. Office and network equipment space is required, and a full-time system administrator/network manager will be needed to monitor the network and help set up and administer other community sites that have purchased connections (such as the public library, local schools, and government offices).

Table 3.3

Level 3 Model—Purchase Equipment and Provide Technical Staff

Startup Costs		
Item	**Who**	**Cost**
Initial design of community site	Web design firm or local help	$500–$2,500
Web server/mail server	N/A	$1,500–$5,000
Network interface equipment	N/A	$1,500–$5,000
Network/Internet feed installation	N/A	$300–$1,500
Software licenses and server software	N/A	$500–$5,000
Ongoing Costs		
Item	**Who**	**Cost (Monthly)**
Community Web site	Community network	N/A
E-mail mailing lists (2–3)	Community network	N/A
Content maintenance	Part-time staff	$500–$2,000
System maintenance and support	Staff or contractor	$500–$2,000
Education/tech support for residents	Staff	$500–$2,000
Offie space and utilities	Per staff member	$300–$500
Internet feed	Telco or IAP	$500–$2,500

Table 3.4 describes the costs associated with operating a level-4 community network.

An interesting thing about this model is that it is possible to break even or make money if enough organizations buy direct connections. By making Internet feeds available to local businesses and property owners at market rates, a community network should be able to become self-financing.

Why direct connections are important

The original Internet design was intended primarily for direct, dedicated connections. In 1990, when the BEV project began, setting up modem access to the Internet required considerable persistence and a fair amount of technical knowledge. As Internet access became more desirable,

Table 3.4
Level 4 Model—Purchase Equipment and Provide Technical Staff

Startup Costs		
Item	**Who**	**Cost**
Initial design of community site	Web design firm or local help	$500–$5,000
Web server/mail server	N/A	$1,500–$5,000
Network interface equipment	N/A	$1,500–$5,000
Multiport router for local Internet feeds	N/A	$10,000–$20,000
Network/Internet feed installation	N/A	$1,000–$2,000
Software licenses and server software	N/A	$500–$5,000
Ongoing Costs		
Item	**Who**	**Cost (Monthly)**
Community Web site	Community network	N/A
E-mail mailing lists (2–3)	Community network	N/A
Content maintenance	Part-time staff	$500–$2,000
System maintenance and support	Staff or contractor	$1,000–$3,000
Education/tech support for residents	Staff	$500–$2,000
Office space and utilities	Per staff member	$300–$500
Internet feed	Telco or IAP	$1,500–$2,500

modem access to the Internet was slowly improved to the point that nonprofessionals could actually get it to work.

Modems transmit Internet data very slowly compared to direct connections fed by T1 lines or other dedicated data services. In addition, modem connections use more software and more network equipment; thus, more things can go wrong. Table 3.5 summarizes some of the differences between modems and direct connections.

A building that is wired for a local (Ethernet) network and a full-time Internet connection gives its owners the option of purchasing an Internet feed and related services from a variety of local, regional, or national Internet providers. This investment in infrastructure (the network equipment and the wiring in the building) is worthwhile because it is relatively easy to change Internet providers as prices and services

Table 3.5
Comparison of Modem and Direct Internet Access

Modems	Direct Connections
Slow	Typically 10 to 100 times faster than modems
Economical only for a few users	As users increase, per user cost decreases
Difficult to set up	Easy to set up
Connections are not very reliable	Connections are very reliable
Intermittent connection to the Internet	Full-time connection to the Internet
It can be difficult to find problems	It is easier to find problems

change—so the cost of hooking users up to the network should decrease over time. The deregulation of the telecommunications industry in early 1996 will hasten competition and directly benefit users of direct Internet connections. Tables 3.6, 3.7, and 3.8 show the cost benefits of using direct connections. Luke Ward's chapter on network technology discusses some of the options for direct connections in much greater detail.

Network administration

Network administration is a critical function of a community network. Once the cables and wires are in place, the routers and servers have been connected to the cables, and end users have their computers plugged in to the network via a modem pool or by direct connection, someone must manage this new community of users in cyberspace.

The level of effort required to do this can be explained most easily by considering the size and scope of your community network.

A level-1 network, with Web space leased from a Web service company or provided by a local nonprofit with an Internet feed (a school, library, or college), requires virtually no network management. What little network management is needed will be provided by the organization that provides your Internet feed. IP address setup and domain name registration are simple because someone else will do it for you.

A level-2 network, in which the community network owns the computer on which a Web server and other services run, also requires little or

Table 3.6
Hooking up 20 Computers in a Local School or Library to the Internet

Solution A: Use Modems to Connect 20 Machines	
Capital costs	20 modems (28,000 bps) × $150 = $3,000
	(Installation of) 20 phone lines × $50 = $1,000
	Cost per connection: $4,000 ÷ 20 = $200 per machine
Ongoing costs	20 phone lines × $16 per month = $320 per month
	20 dialup Internet accounts × $20 = $400
	Monthly cost per connection: $720 ÷ 20 = $36
	Price/performance: $36 ÷ 28,800 = 0.1250
Solution B: Use Direct Connection (T1 Line) to Connect 20 Machines	
Capital costs	20 Ethernet cards (1.5 million bps) × $75 = $1,500
	Ethernet hub and wiring: about $800
	CSU/DSU and router for T1 line: about $3,000
	Installation of T1 line: about $500
	Cost per connection: $6,800 ÷ 20 = $340 per machine
Ongoing costs	1 T1 line with Internet feed: $2,500 per month
	Monthly cost per connection: $2,500 ÷ 20 = $125
	Price/performance: $125 ÷ 1,500,000 = 0.0083

no network administration. Once the computer has been set up and configured properly, there are few network administration tasks that need to be done.

A level-3 network, with multiple computers hosting several different kinds of Internet services, can be more complex, but not overly so. Whoever manages the Web, news, and mail servers can also handle such issues as virtual domains for groups that want their own Web address (e.g. http://boy-scouts.ourtown.va.us/). As with level-1 and level-2 configurations, the organization providing your Internet feed will handle most network configuration issues for you.

A level-4 network, in which the community network provides Internet connections to other organizations through direct connections and/or a modem pool, is very different. Now, the community network is not only a user of IP (Internet Protocol) computer addresses, but it is also a supplier. Each computer hooked to your Internet feed must use an IP

Table 3.7

Hooking up 100 Computers in a Local School or Library to the Internet

Solution A: Use Modems to Connect 100 Machines	
Capital costs	100 modems (28,000 bps) × $150 = $15,000
	(Installation of) 100 phone lines × $50 = $5,000
	Cost per connection: $20,000 ÷ 100 = $200 per machine
Ongoing costs	100 phone lines × $16 per month = $1,600 per month
	100 dialup Internet accounts × $20 = $2,000
	Monthly cost per connection: $3,600 ÷ 100 = $36
	Price/performance: $36 ÷ 28,800 = 0.0556
Solution B: Use Direct Connection (T1 Line) to Connect 100 Machines	
Capital costs	100 Ethernet cards (1.5 million bps) × $75 = $1,500
	Ethernet hub and wiring: about $800
	CSU/DSU and router for T1 line: about $3,000
	Installation of T1 line: about $500
	Cost per connection: $14,000 ÷ 100 = $140 per machine
Ongoing costs	1 T1 line with Internet feed: $2,500 per month
	Monthly cost per connection: $2,500 ÷ 100 = $25
	Price/performance: $25 ÷ 1,500,000 = 0.0017

address, and the administration of these addresses must be handled by the community network. The supplier of your Internet feed will provide a block of IP addresses (of the form xxx.yyy.zzz.aaa; for example, 128.173.28.92). Each computer on the network will have one of these Internet "street addresses" as well as an English language equivalent. A DNS[4] is required to convert the English language addresses into their numeric equivalents.

If your network is very small (say less than 200 addresses), the organization providing your Internet feed may be willing to provide DNS services for you. If you have several hundred computers hooked to your network, you will probably have to run your own "primary" DNS server,

---

4. For more information on DNSs read the early part of this chapter and/or Chapter 10.

Table 3.8
Hooking up 200 Computers in a Local School or Library to the Internet

Solution A: Use Modems to Connect 200 Machines	
Capital costs	200 modems (28,000 bps) × $150 = $30,000
	(Installation of) 200 phone lines × $50 = $10,000
	Cost per connection: $40,000 ÷ 200 = $200 per machine
Ongoing costs	200 phone lines × $16 per month = $3,200 per month
	200 dialup Internet accounts × $20 = $4,000
	Monthly cost per connection: $7,200 ÷ 200 = $36
	Price/performance: $36 ÷ 28,800 = 0.0556

Solution B: Use Direct Connection (T1 Line) to Connect 200 Machines	
Capital costs	200 Ethernet cards (1.5 million bps) × $75 = $3,000
	Ethernet hub and wiring: about $18,000
	CSU/DSU and router for T1 line: about $3,000
	Installation of T1 line: about $500
	Cost per connection: $24,500 ÷ 200 = $122.50 per machine
Ongoing costs	1 T1 line with Internet feed: $2,500 per month
	Monthly cost per connection: $2,500 ÷ 200 = $12.50
	Price/performance: $12.50 ÷ 1,500,000 = 0.0008

and possibly a "secondary" DNS server, which is used if the primary server quits working.

As computers are added and removed from your network, you must allocate and retrieve the IP addresses assigned to them. Some of this can be automated, but people are involved in making the requests. As a result, there will be a significant amount of someone's time that must be dedicated to the task of managing IP addresses.

When other groups and organizations purchase Internet access from you, there will often be a router and hub installed at their building or site. These pieces of equipment must be configured properly, with their own Internet addresses; and issues like security and IP address allocation at that site must be considered. When supplying an Internet connection to a school, library, or office building, there may be no one there who is qualified to manage the equipment; so the community network must provide technical support, which often requires good communications skills as well as good technical skills.

There are at least two ways to ensure that the work is done properly. You can hire someone, or you can contract for help. If you are running a mail server, one or more Web servers, providing other kinds of Internet services, and providing Internet access to other locations, you probably have enough work to keep at least one technical person busy full-time. If the same person must support the information services and support the network, someone with a background in software and programming (and an interest in networks) may be a good choice. If there does not appear to be enough work and/or income from the network to justify keeping someone on the payroll as a regular employee, you may be able to contract with your Internet provider or a local technology consulting firm to do the work for you. Keep in mind that much of the work involves talking to and helping your human customers, which may often be time-consuming. If someone is being paid by the hour to do this, it could become expensive quickly.

In Blacksburg, the BEV group has had the (part-time) help of several network specialists at Virginia Tech in addition to at least one full-time person. The chores that have taken the most work are registering domain names on the BEV DNS for local Web service companies, providing SLIP addresses for modem pool users, and providing technical assistance and network management to remote sites like public schools and apartment complexes with Ethernet.

One of the most difficult issues of network management is finding a person who has the proper qualifications and experience to do the work. Network engineers and support specialists are still relatively rare, and demand in the private sector (in 1999) for such people is much higher than the supply. Starting salaries for engineers with one or two years of experience can easily exceed $40,000. This is an enormous sum for a community network that may rely heavily on donations and grants. It may be easier to find two people to do this work on a part-time basis than to find one full-time person. It may also be necessary to hire someone with very little experience; in this case, he or she should be very bright and eager for on-the-job training.

Once you have someone in your organization performing tasks like running your DNS, managing IP addresses, and configuring routers, he or she will become a very valuable person, not only to you but to any other business or organization that needs a similar kind of help. Expect high turnover in this position. Insist on careful written documentation of all procedures and automated tasks (software), and ensure that there is at least one other person in the organization that can perform basic network management tasks in the absence of your primary support person.

4

Evaluating the Blacksburg Electronic Village

by Scott J. Patterson

THE SUCCESS OF a technological innovation depends largely on the degree to which the innovation meets the needs of the people who use it. Of early interest to the designers of the BEV was concern over how well the system would meet the needs of the BEV members. To facilitate the evaluation effort, BEV designers gathered together a range of experts in computer-mediated communication systems from Virginia Tech. The initial hope was that a multidisciplinary orientation to a BEV research program would capitalize on the strengths of the entire university in understanding the uses and impacts of the BEV.

A meeting was called and attended by BEV system designers, computer scientists, industrial and system engineers, educators, town managers, and social scientists. One theme of that session was the realization of the need for a unified vision of what the BEV is: a model. The first BEV model was a parsimonious one.

That first model consisted of four boxes connected linearly. The first box represented BEV personnel, those persons responsible for constructing and maintaining the system. The second box was the variety of information available in the village. The third box represented the various physical mechanisms by which the village would operate. The fourth box represented the users of the system. The participants in the meeting became enthusiastic about the vision of the model; this was the BEV. The vision of the BEV was represented in a model of communication, or, more accurately a version of Shannon and Weaver's mathematical model of communication [1]. Box 1 is the senders, box 2 the messages, box 3 the channels, and box 4 the receivers.

It was exciting to observe how the members of the advisory panel realized the usefulness of communication theory in creating a shared vision of an electronic village. That excitement was tempered, however, by the understanding that communication theory had advanced quite a bit since 1949. While simple and parsimonious, the Shannon and Weaver model does not capture all the nuances of what it means to communicate. An explication of the model should explain how advances in communication theory could both provide the unifying vision of the village discovered by this group and meet the group goal of helping the BEV serve its community.

The purpose of this chapter is to explicate a model that captures the primary realms that we must study to evaluate the BEV. The first section outlines the underlying assumptions about the purposes of evaluation. The second section develops a model for evaluation of community computer communication systems. The final section details the specific research activities done in the BEV evaluation and integrates them with the evaluation model.

The assumptions of an evaluation system

Three primary assumptions underlie the evaluation of the BEV: the social construction of technology, the interpenetration of use and design, and the importance of a multidisciplinary approach to evaluation. This section unpacks these three assumptions.

The general purpose of the evaluation is to enable the innovation to be responsive to the needs of the people who use it. The assumptions explicated here facilitate the role of evaluation in providing the feedback important to catalyze system adaptability. Thus, the model providing a picture of how to evaluate the BEV or any BEV-like initiative needs to be robust enough to incorporate these three assumptions.

The BEV as socially constructed technology

Once a technology begins to diffuse throughout a culture, the culture begins to shape and mold the technology just as the technology transforms the culture. This premise is basic to the social constructivist view of technology [2, 3], which seeks to understand how technology is shaped by social factors. Key to a social constructivist understanding of technology is how people come to use a technology. Schwartz-Cowan points to the "consumption-junction," that is, "the place and time at which the consumer makes choices between competing technologies," as central to an understanding of a given technology's diffusion [4]. This focus on the user/consumer in the social construction of technology has been reiterated by other social constructivists [3, 5, 6].

The history of another communications medium, the telephone, provides a classic example of how technology is defined by its use. For over 100 years, observers have speculated about the potential household uses of the telephone. DuMoncel outlined five "applications for the telephone," including domestic applications such as giving orders to porters and servants [7]. Ten years later, Preece and Maier reiterated DuMoncel's typology, coming to the additional conclusion that while the telephone is useful for short-distance transmission of sound, it is unfeasible for transmission over long distances [8]. The telegraph was still thought to be the only effective means of distance communication.

Those early writers did not foresee that users of the telephone would take over the diffusion process and co-opt that new technology to meet their own needs. Indeed, as early as 1890, telephone company executives used the technology to transmit a live concert from the concert hall to their homes for the purposes of a private party [9]. Since that time, the telephone has been adapted to meet a range of human communication needs, displacing the supposedly superior telegraph technology as a preferred means of long-distance communication. Indeed, the realization by AT&T in the 1920s that the strongest market for telephone services were women talking in their homes to other women led Theordore Vail to realign AT&T's marketing efforts away from an emphasis solely on business telephone service toward a campaign targeted at the sociability uses of the telephone—"Reach Out and Touch Someone."

The primary assumption underlying evaluation of the BEV is that understanding how people use the system to meet their needs is central to the evaluation effort. Research on phones justifies the expectation that the BEV is going to become what the people who use it want or need it to be.

The interpenetration of design and use

In addition to the assumption that a technological system is defined by how people use the system, the evaluation model must also realize that, to a great extent, the range of available uses is constrained and liberated by the design of the system. Returning to the telephone example, the only reason that the telephone was used for sociability is because it was already in the home. Phones entered the home because the telephone company encouraged businessmen to have a phone at home in case they were needed by the office after normal working hours or to "call home and tell your wife that you would be bringing a guest home for dinner" [10]. The designers of the early telephone system introduced the phone as a home appliance on the assumption that it would be used only for business purposes. Would the telephone have become an appliance for sociability if it had never been designed for home use?

In addition to a system changing to adapt to how it is used, the designers of the system must make presumptions about how it *might* be used. Decisions ranging from the design of the user interface, to the types of services included, to the price, to the location of public access sites all affect how the system is used. In addition to social constructivism, the evaluation model needs to incorporate the presumption that any given communication system is determined, to a great extent, by the initial expectations and efforts of the system designers. The evaluation model must be guided by the presumption that design and use interpenetrate in the process of diffusing the technology.

A multidisciplinary approach

The final assumption underlying the model of evaluation developed here is that the model must encourage a multidisciplinary approach to the study of the BEV or any other community communication initiative. The intuitive appeal of such inclusiveness is that the more eyes you have watching something, the more accurate picture you are able to paint regarding what happened. This is the basic assumption of triangulation in social science research [11, 12]. Different scholars and experts will have different skills, theories, and techniques they can bring to bear on the substantive phenomena under investigation.

The model of evaluation proposed here is targeted toward the same substantive domain, namely, community computer communication initiatives—the constant in the equation. The two variables involved in the equation are the conceptual and the methodological domains. The conceptual domain involves the application of specific theories to an understanding of the substantive domain. The importance of a multidisciplinary approach is easily evidenced in an analysis of the conceptual domain. Different disciplines or approaches to the phenomena of community computer communication systems have different theories that can help explain what is happening. Sociologists can bring theories of community to bear, education scholars have theories that describe the impact of BEV-like systems on educational institutions, and political scientists can similarly understand the impact of a technological innovation on political institutions. These examples are just from the social sciences. Researchers from the hard sciences, engineering, urban planning, theater, anthropology, literature, and almost any other discipline all have some important conceptualizations to bring to the evaluation of community computer communication initiatives. An effective model of evaluation should provide a vision that researchers from a variety of disciplines can plug into.

The methodological domain is also aided by the incorporation of researchers from across the intellectual spectrum. The methodological domain deals with the tools that researchers use to see empirically what is happening. Reliance on a single method of making observations is a recipe for a myopic vision of the substantive domain. An ideal evaluation program should incorporate the diversity of methodological approaches available, including survey research, experimental design, ethnography, and historical approaches. The evaluation system should encourage both qualitative and quantitative observation.

Evaluation of communication initiatives such as the BEV need to be multidisciplinary in nature, that is, seen from multiple perspectives. The substantive realm remains relatively constant. The evaluation always concerns the same phenomenon, the communication initiative. However, the validity of the evaluation increases as experts from many different disciplines bring their discipline-specific conceptualizations and methodologies to bear on the same problem. The model guiding these diverse investigations must provide for the importance of a multiple-perspective approach.

A model for the evaluation of the BEV

This section explicates a model for the evaluation of community communication initiatives like the BEV. The development of the model proceeds along three steps. First, the importance of models in providing a framework for a valid feedback system is discussed. Second, a variety of approaches to modeling communication behavior are explored as a starting point for the development of a general model. Finally, a generalized model of evaluation is proposed and discussed.

The importance of models in evaluation

Deutsch indicates that models are useful in providing simple explanations of complex phenomena [13]. This function was manifest in the first BEV planning meeting, in which the Shannon and Weaver model was proposed to provide a common understanding of a complex communication system. Plax suggests that consultants use a specific research model as a baseline or framework from which to understand the specific evaluation problem [14]. He also argues that the evaluator's role as applied researcher should be "informed by formal theory" [14]. Nolan also argues that the successful consultant is "one who conducts communication inquiry within the boundaries and definitions of social science research" [15].

Clearly, any scholar attempting to evaluate should be guided by theory. This guidance informs the ability of the consultant to understand the phenomena under examination. The ability to model this theory is important for three reasons. First, the model provides the researcher with a parsimonious conceptualization of the problem under investigation. The model helps direct the evaluation process. Second, a model provides the evaluation researcher a vehicle with which to structure the communication of evaluation outcomes to the audience. For example, if all the researchers participating in the evaluation process use the same model to structure their presentation of the evaluation results, the audience for these varied presentations (e.g., BEV system designers) is provided with a consistent structure, which facilitates understanding. The model provides a shared vocabulary. Finally, agreement on a common model or vision of the evaluation target facilitates communication among the evaluators. A common model can aid in the translation of the conceptual and methodological vocabulary of diverse disciplines. Thus, a model eventually serves to facilitate the evaluation process.

Models of communication behavior

The BEV is a community communication initiative. Thus, a logical place to begin looking for a model of the BEV is in the communication literature. Indeed, beginning with communication models is logical, given the first generally agreed on model of the BEV was a communication model.

McQuail and Windahl describe three general categories of mediated communication models: transmission models, ritual models, and attention models [16]. Transmission models, represented by early models of communication, identify the components of a communication system and discuss how messages and meanings flow through those systems. The Shannon and Weaver model is representative of a transmission model. Ritual models are concerned with communication as a social force. They attempt to explain how messages and meanings create and change the cultures, beliefs, and values of society. Attention models are concerned with an understanding of the media's economic goal of gaining and maintaining audience attention. McQuail and Windahl argue that communi- cation inquiry is successfully guided from within any of these three categories.

While models of mediated communication fall into three broad categories, McQuail and Windahl subdivide models of communication in the information age into three areas: (1) general models of information society, (2) models that describe the flow of information traffic, and (3) models that attempt to provide a vision of the regulatory environment surrounding the creation of the information technology and business.

None of these existing models of communication in the information age provided a good conceptual fit for the BEV, for two reasons. First, none of the models captured the phenomenon of a grassroots, community-sponsored information society. Most of the models focus on either the national/global level or the level of the organization (e.g., models of information technology in manufacturing organizations). Second, most existing models that did apply to the BEV were complex. They tried to capture the range of communication influence extant in a new communication system. The evaluation of the BEV did not need a model of communication, but rather a communication-based model of the electronic village concept. BEV needed a model as simple as Shannon and Weaver's model, one that captured the essence of the range of communication theory.

A Tetrad model of evaluation of community networks

A tetrad model of evaluation of community networks is shown in Figure 4.1. The model is patterned after McLuhan and Power's model of

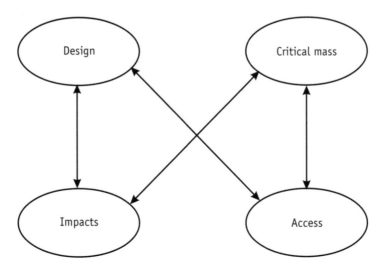

Figure 4.1   A model for the evaluation of a community computer network.

transformation in the global village [17]. The model consists of four nodes connected via a tetrad. The nodes are design, access, critical mass, and impacts.

The design node relates to the processes involved in the construction and operation of a communication initiative. Those processes include user interface design, system architecture, physical connections, capacity, hardware, user support, access modes, and feedback mechanisms.

Access refers to how accessible the system is to the users. Following Hudson, access is understood as a function of four factors: cost, physical distance from an access point, quality of service, and user knowledge and expectations [18].

Critical mass describes the process by which the system becomes sustainable, or the point that universal access is achieved [19, 20]. Critical mass asks who is using the system. The value of the application of critical mass theory is twofold. First, it provides a picture of the diffusion of interactive communication systems. That picture provides a referent for all scholars as to the current status of diffusion. The needs of the system and its users are different at different stages in the diffusion process. Second, while not in complete agreement on this point, critical mass theories posit that an interactive communication initiative reaches critical mass, or is sustainable, when 19% of the population use the system [21]. Thus, the critical mass point becomes a marker by which to situate the longitudinal evaluation of a BEV-like system. An interactive communication system is more likely to reach critical mass when high-interest and high-resource individuals are available on the network. Measuring the number of users, amount of use, types of use, number of service providers, and types of services helps in determining critical mass.

Impacts are the consequences, intended or not, of use. The impacts can be at the individual level (e.g., psychological gratification), the organizational level (e.g., education, small business, government), the social level (e.g., community involvement), and the cultural level (e.g., economic structures).

The tetrad represents a synchronic and diachronic process that connects the nodes in a system of relationships. A tetradic structure is proposed for this system rather than a cat's-cradle structure in which every node is connected to every other node. The tetrad represents the notion of the interdependency of the nodes. Each node in the model is equally important. The tetrad also captures the notion that no single node has all the explanatory power in the model. Each node is equally affected by two other nodes. For example, an understanding of critical mass is informed by its relation to access (What are the characteristics of the people using the system? Are they high-resource users?) and to impacts (Does the critical mass of users perceive that it is affecting individuals, organizations, the community, and the culture?). Each node has one other node that it can affect only through the moderating impact of the other node. For example, in this model it is impossible for the system design node to directly affect critical mass. The effect of design on system sustainability (critical mass) is moderated by both access (Is the system easy to use? Does the system break down? Is the cost reasonable? Are there sufficient public access sites?) and impact (Are the services provided by the system gratifying to the users? Does the system encourage greater civic activism? Do parents who join become part of the critical mass because of the educational structures designed into the system?).

Research activities in the four nodes

This section explicates a large number of ongoing research activities being conducted as part of a large-scale evaluation of the BEV. The utility of the BEV evaluation model is evidenced in its ability to organize each of these research activities.

The evaluation of the BEV represents efforts from a variety of disciplines, from engineering to education to communication to anthropology. The BEV research home page (www.bev.net/project/research) identifies 45 scholars from 12 disciplines who are involved with research projects related to the BEV. The experts from these disciplines use a wide range of research techniques, from archival methods to surveys to ethnography to experimental design. The evaluation of the BEV is both multiconceptual and multimethodological. The BEV has a full-time director of research who mobilizes faculty, directs and coordinates research areas, holds seminars to showcase work in progress, and pursues grants and other funds to support this large-scale research effort.

To better describe the utility of the proposed model in classifying the disparate evaluation activities conducted in relation to the BEV project, this section is organized by node.

Design

Evaluation efforts in the design node have centered around three primary activities: the BEV history project, evaluation of interface design, and evaluation of structures designed to enhance K–12 education via the BEV.

BEV history project

The BEV history project led by Jack Carroll of the computer science department at Virginia Tech has four long-term goals:

- Identify the main issues and activities relating to community networking projects;

- Develop a digital history and ethnography of the BEV;

- Construct a community-oriented digital library testbed for the development of automatic, self-sustaining, usable, intelligent hypermedia services;

- Elaborate a model for successful community networks that can be easily applied.

The current results of the BEV history project are discussed elsewhere in this book and are available online at history.bev.net/bevhist/.

Interface design

Early research conducted into the evaluation of the BEV system software package was conducted via focus group interviews with beta testers of the early BEV software package [22]. Those early focus groups discovered that the BEV package needed to be easier to install and that the software needed to be available on a variety of platforms. Also, as a part of the software design and installation process, the focus groups also indicated the importance of user support in the startup stages for new users.

Further research into design of the BEV user interface can be found on the BEV research page (crusher.bev.net/project/research/) and also on the Project Succeed home page (fiddle.ee.vt.edu/succeed/ home.html) of the Department of Electrical Engineering at Virginia Tech. These projects on enhancing the design of computer communication systems all plug into the BEV. Project Succeed and the Electronic Connectivity Group have been especially active in exploring how technologies of desktop video conferencing, groupware, and the WWW can be used to help users performing tasks related to information dissemination, counseling and mentoring, research collaboration, and distance learning. Evaluation of the usability of these communication technologies is being carried out by the Virginia Tech Human Factors Engineering Center.

Curriculum design

A final area where the system design node has been extensively researched is in relation to developing curriculum for the K–12 schools in Blacksburg and in Montgomery County. The K–12 curricular initiative focuses specifically on using BEV and Internet communication technologies to create virtual collaborative classrooms. These virtual classrooms plan to integrate the resources of the entire community (libraries, homes, businesses, government) and global networks into the educational experience. The details of the design and evaluation efforts related to educational organizations in Blacksburg can be found on the virtual school's Web page (http://crusher.bev.net/project/research/planning.html).

Access

Research into the accessibility of the BEV has been carried out along two fronts. First, in collaboration with the Town of Blacksburg, BEV evaluators have conducted yearly mail surveys of the Blacksburg community about their BEV knowledge, experience, and attitudes. The details of these research efforts can be accessed via the BEV research page. The general conclusions of those surveys in relation to the question of the accessibility of the BEV have been encouraging. The first two years' surveys indicate that the majority of the residents of Blacksburg have little geographic distance to overcome to obtain BEV access. The majority of the respondents indicate they have access to a computer at home or at work. The general level of computer knowledge and experience is also high in Blacksburg. Finally, Patterson and Kavanaugh conclude that the residents of Blacksburg generally have diverse and high expectations as to what the BEV can do for them [23].

The second way access has been evaluated in the BEV is through an analysis of the transaction records of the BEV project servers. These statistics are kept for the BEV WWW server, the home of the BEV. Statistics have shown a general pattern of increasing access since the BEV's inception.

Critical mass

The critical mass node has been studied primarily through the use of a user-profile questionnaire developed by Patterson, Bishop, and Kavanaugh [22]. When new users register for the BEV, they are asked to complete a user profile, which asks for information about their experience with computers and computer-mediated communication, their expectations for the BEV and the Internet in general, and a variety of demographic indicators. In a study based on those surveys, Patterson and Kavanaugh conclude that the BEV has reached a point of critical mass, or general system sustainability [24].

Impacts

Research into the impacts of the BEV have centered on three primary realms: individuals, organizations, and community. At the individual level, Patterson has conducted telephone surveys of BEV users in an attempt to measure the psychological gratifications they seek from the BEV [25]. The results of this research indicate that users derive gratification from the BEV related to local news and information, entertainment, health care information, and computer knowledge and experience. In an ongoing project, Patterson is investigating the impact of BEV communication technology on the use of traditional communication technologies such as the telephone and television. In that study, Patterson uses a combination of focused interviews and diaries as data collection techniques. The impact of the BEV on traditional media has already been confirmed by Bromley and Bowles [26]. An electronic survey of BEV users concluded that the BEV may have substantial impacts on the use of traditional newspaper media.

At the organizational level, Patterson and Kavanaugh have conducted mail surveys of business users and nonusers [27]. The K–12 curriculum design group is conducting an ongoing ethnography that captures the successes, failures, and developments of the project. The county libraries, primary BEV public access sites, are also conducting ongoing research on the relation between the BEV and the community library system.

At the community level, a current research project is attempting to examine the market for BEV services. As part of that study, the project is comparing users and nonusers across a variety of community variables, including community attachment and involvement. One hypothesis of the study comparing users and nonusers is that users of the BEV will demonstrate higher levels of community attachment and involvement than nonusers. The project uses an online survey of BEV users and telephone survey interviews with nonusers as data collection techniques.

Lessons learned

What lessons have been learned from the evaluation efforts of the BEV? First, interdisciplinary collaboration is key in obtaining a valid picture of a community computer network. No single discipline has the ability, conceptually and methodologically, to create a robust picture of the substantive phenomena. New electronic villages should strive to recruit expertise from across the intellectual milieu at the beginning of their evaluations.

Second, it is essential to begin the evaluation process prior to the start of the technological intervention. Here in Blacksburg, focus group interviews with innovators, interviews with system designers, and a general survey of the community were conducted prior to system implementation. This data is the baseline to which further data points can be compared. The evaluation needs to be conceptualized longitudinally.

Third, it is essential that all members of the evaluation team agree to the public dissemination of their results and findings. Public dissemination can be facilitated by the use of a common model as outlined in this chapter. The model outlined here serves as an organizing structure for the analysis of the evaluation efforts and aids in the communication of those efforts to other evaluators, system designers, members of the system, and designers of future community networking initiatives. Evaluation research of the BEV is distributed online, via BEV research talks, professional organizations, scholarly journals, and media outlets. It is essential that future electronic villages make the results of their evaluations, both good and bad, available to the public.

Fourth, it is important to extend the evaluation process by comparing the data from one community to that of another community. In essence, the evaluators must raise the unit of analysis to the level of the community. Not until evaluators are able to compare different cases are they truly able to create a general theory of community computer networking that can be generalized to all such initiatives. Public dissemination of the data and the results of evaluation efforts is central to the facilitation of this comparative work.

Finally, it should be remembered that the purpose of the evaluation is to empower the target of the evaluation. The initial premise of this chapter is that the BEV needs to be responsive to the needs of the people who use it. Evaluation is the feedback mechanism by which the BEV can learn how to adapt. Further, the BEV system has incorporated the results of the evaluation research into real changes in the BEV. This communication between evaluators and designers is essential for the success of the community networking initiative.

References

[1] Shannon, C., and W. Weaver, *The Mathematical Theory of Communication,* Urbana, IL: University of Illinois Press, 1949.

[2] Pinch, T., and W. Bijker, "The Social Construction of Facts and Artifacts: Or How the Sociology of Science and the Sociology of Technology Might Benefit Each Other," in W. Bijker, T. Hughes, and T. Pinch (eds.), *The Social Construction of Technological Systems*, Cambridge, MA: MIT Press, 1984.

[3] Dierkes, M., and U. Hoffman, "Understanding Technological Development as a Social Process: An Introductory Note," in M. Dierkes and U. Hoffman (eds.), *New Technology at the Outset: Social Forces in the Shaping of Technological Innovations*, Boulder, CO: Westview Press, 1992.

[4] Schwartz-Cowan, R., "The Consumption Junction: A Proposal for Research Strategies in the Sociology of Technology," in W. Bijker, T. Hughes, and T. Pinch (eds.), *The Social Construction of Technological Systems*, Cambridge, MA: MIT Press, 1987.

[5] Dankbaar, B., and R. van Tulder, "The Influence of Users in Standardization: The Case of MAP," in M. Dierkes and U. Hoffman (eds.), *New Technology at the Outset: Social Forces in the Shaping of Technological Innovations*, Boulder, CO: Westview Press, 1992.

[6] Schmidt, S., and R. Werle, "The Development of Compatibility Standards in Telecommunications: Conceptual Framework and Theoretical Perspective," in M. Dierkes and U. Hoffman (eds.), *New Technology at the Outset: Social Forces in the Shaping of Technological Innovations*, Boulder, CO: Westview Press, 1992.

[7] DuMoncel, C., *The Telephone, the Microphone, and the Phonograph*, London: Kegan Paul, 1879.

[8] Preece, W., and J. Maier, *The Telephone*, London: Whittaker & Co., 1889.

[9] Barnouw, E., *A History of Broadcasting in the United States: A Tower in Babel*, New York: Oxford, 1966.

[10] Fischer, C. S., *America Calling: A Social History of the Telephone to 1940*, Berkeley: University of California Press, 1992.

[11] Babbie, E., *The Practice of Social Research*, New York: Wadsworth, 1992.

[12] Brinberg, D., and J. McGrath, *Validity and the Research Process*, Newbury Park, CA: Sage, 1985.

[13] Deutsch, K., *The Nerves of Government*, New York: Free Press, 1966.

[14] Plax, T. G., "Understanding Applied Communication Inquiry: Researcher as Organizational Consultant," *Journal of Applied Communication Research*, 1991, pp. 55–60.

[15] Nolan, L. L., *Communication Intervention: What Do Organizations Want?* Paper presented at the annual meeting of the Speech Communication Association, Atlanta, GA, November 1991.

[16] McQuail, D., and S. Windahl, *Communication Models for the Study of Mass Communication*, 2d Ed., New York: Longman, 1993.

[17] McLuhan, M., and B. Powers, *The Global Village: Transformations in World Life and Media in the 21st Century*, New York: Oxford, 1989.

[18] Hudson, H., *When Telephones Reach the Village: The Role of Telecommunications in Rural Development*, Norwood, MA: Ablex, 1984.

[19] Rogers, E., "The 'Critical Mass' in the Diffusion of Technologies," in M. Carnevale, M. Lucertini, and S. Nicosia (eds.), *Modelling the Innovation: Communications, Automation, and Information Systems,* North-Holland: Elsevier Science Publishers B. V., 1990.

[20] Valente, T., *Network Models of the Diffusion of Innovations*, New York: Hampton Press, 1995.

[21] Markus, M. L., "Toward a 'Critical Mass' Theory of Interactive Media: Universal Access, Interdependence and Diffusion," *Communication Research*, 14(5), 1987, pp.491–511.

[22] Patterson, S., A. Bishop, and A. Kavanaugh, *Preliminary Evaluation of the Blacksburg Electronic Village, Grant Report,* Council on Library Resources, Washington, D.C., 1994.

[23] Patterson, S., and A. Kavanaugh, *Rural Users' Expectations of the Information Superhighway*, Media Information Australia, 1995.

[24] Patterson, S., and A. Kavanaugh, *Building Critical Mass in a Community Network,* manuscript submitted for publication, 1996.

[25] Patterson, S., *Gratifications Sought From the Electronic Village*, paper presented to the Annual Conference of the International Communication Association, Sydney, Australia, May 1994.

[26] Bromley, R., and D. Bowles, "The Impact of Internet on use of Traditional News Media," *Newspaper Research Journal*, 16(2), 1995, pp. 14–27.

[27] Patterson, S., and A. Kavanaugh, *The Impact of a Community Computer Network on Business: Comparisons of Business Users and Non-Users,* Unpublished manuscript, Virginia Polytechnic Institute and State University, Department of Communication Studies, Blacksburg, 1996.

5

The Use and Impact of the Blacksburg Electronic Village

by Andrea Kavanaugh, Andrew Cohill and Scott Patterson[1]

RESEARCH AND EVALUATION of the use and impact of computer networking in Blacksburg, Montgomery County and environs indicates that it is being used to access information more conveniently and to stay in communication with friends, family, and cohorts regarding the news, activities, and logistics of their daily lives. Growth in computer and network use has been steady over the five-year period since the project began in the fall of 1993. The majority (86%) of residents in the town of Blacksburg report using the Internet (January 1999), and about one-fifth of the non-Blacksburg population of Montgomery County was using the Internet. As with other parts of the country, in Blacksburg and environs, e-mail and the WWW are the most popular services.

---

1. We would like to acknowledge the significant contribution of research assistants Armando Borja, Sarah Laughon, Dean Orrell, and Michael Mosteller.

Online communication, particularly through Web sites and group discussion forums such as listservs, are excellent strategies for organizations to extend their missions, develop new partnerships, and expand existing relationships. By offering a package of online services to existing community organizations and groups, a community network can stimulate communication and strengthen social networks. The BEV "Community Connections" package (consisting of a Web site, listserv, and two e-mail accounts for $20 a year) facilitates communication among group members. There are numerous organizations and informal groups, such as the local school board, the high school band, boy scout troops, the soccer league, the PTA, and churches, that have been using Web sites and listservs very effectively toward these ends. Users report that convenient access to timely information and updates has been a major improvement over traditional communication mechanisms such as telephone trees. Community leaders find it much easier to reach their constituents with organizational information (such as minutes, agendas, and background documentation) and to discuss issues. In many cases, without online outlets, these documents would not get distributed at all. The increased distribution of background information and discussion among constituents increases group members' sense of involvement in issues and strengthens their association with the organization and its membership.

These effects are common where individuals have a pre-existing need to communicate with each other due to membership in the same organization or group. These results are much less likely to occur where the listserv is broadly defined either by topic, such as "education" or "environment," or by users, such as the community in general. It appears that trying to create online "communities" out of thin air (i.e., without some pre-existing reason to come together) is not an effective strategy. Rather, the Internet seems to reinforce and strengthen community where it already exists. This may explain in part the results of the infamous Carnegie-Mellon study in 1998 that "showed" that the Internet increases isolation. The study may have been fatally flawed because participants in the study were selected without regard to any pre-existing community relationships. That is, it seems likely that the participants in the CMU study were lonely and isolated before they began using the Internet—the Internet did not cause them to become lonely. This is an important issue to consider as communities debate why they should invest in the financial support of a community network project. One important reason to use tax dollars or other broad funding bases for a community network is because it appears that ties among groups and individuals in the existing community are strengthened.

For civic engagement, on the other hand, topics and issues are often diffuse and constituents broadly dispersed. Involvement occurs through a variety of means and organizations. In this context, involvement in local community is more likely to increase with Internet access among people who are predisposed to be involved in the community. Among people who are not "poised to be active," Internet access does not increase involvement.

The essential purpose and meaning of a community network is reinforced through a sense of local ownership. This requires that members of a community have acquired necessary skills to build local content through training and user support. Training specialized user groups (such as community leaders, school teachers, businesses, elected officials) as well as the general public can be accomplished through various organizational outlets and programs. These include the local public library, the YMCA, and other community nonprofit organizations; private companies specializing in business training; the public school system; and town recreation programs. Transferring skills and knowledge throughout a broad, diverse population in the community assures local ownership of online content while building a critical mass of users. Local ownership of the community network is another key issue currently under debate as many communities are receiving offers from commercial firms offering to provide "free" community network systems in exchange for allowing these companies to put banner advertising on the site. Advertising issues aside, a company in Seattle, Washington, trying to manage a community network in North Carolina is not likely to be able to adjust programs and services to the special and unique needs of that North Carolina community. Do we really want to take a Wal-Mart or McDonald's approach to the creation of our online communities? By handing over online communities to commercial "one size fits all" systems, we are really saying that there is nothing special or different about our community from any other. Is this what we really want?

It is important for community network designers and managers to focus on people and content, not technology. Technology, such as network connections and software, can be readily provided by private companies. Local content is best developed by local users who understand local needs. Mobilizing people and meeting local needs of diverse users requires human judgment and tactful, timely intervention. The evaluation of communication behavior and effects will help designers and managers intervene appropriately, and as needed.

This chapter elaborates on these highlights of findings based on a series of quantitative and qualitative studies conducted since 1994 with support from the Town of Blacksburg, the Virginia Tech Center for Survey Research, the Virginia State Library Association, the U.S. Department of Commerce (TIIAP), and the National Science Foundation.

Profiles of users

The majority (85%) of Internet users in Blacksburg is affiliated with Virginia Tech; the majority of Tech affiliates are students. Most Internet users in Blacksburg are either employed as professionals or identify themselves as full-time students. Almost half (48%) of the users accessing the Internet free of charge through the local public library (Montgomery Floyd Regional Library) are female.

Outside of Blacksburg, an estimated 18–20% of Montgomery County residents use the Internet. These users represent diverse backgrounds (farming, retail, university faculty, and staff). Several of this chapter's tables show differences in responses for services between Blacksburg residents and non-Blacksburg residents of Montgomery County.

Early trends on user expectations have been captured with a user profile questionnaire, completed when individuals signed up for an Internet account through Virginia Tech between 1994 and 1996. After 1996, private ISPs also began to offer access, so it was no longer possible for BEV to obtain completed questionnaires from subscribers as they signed up for service. In 1997, only individuals signing up for apartment Ethernet service came to the BEV Office, at which point they completed the user profile questionnaire. The apartment Ethernet subscribers, who obtained a high-capacity Internet connection (up to 10-Mbps), were most often students at Virginia Tech. In a 1997 survey mailed to all subscribers signed up between 1994 and 1996, subscribers reported trends in their actual Internet use (Patterson and Kavanaugh, 1997).

Use and expectations of the Internet

Table 5.1 shows early trends in subscriber expectations during the 1994–96 period as well as for the "Ethernet-only" users in Blacksburg apartments (1997). The table shows the percentage of respondents on subscriber questionnaires reporting they expect the Internet to be "somewhat" or "very helpful," the two highest points on a four-point scale). Figures 5.1 and 5.2 chart the findings presented in Table 5.1.

E-mail and the WWW are the two most popular online services. The percentage of users reporting they are "somewhat" or "very interested" in e-mail is 97%; similarly, the percentage of users reporting that they are "somewhat" or "very interested" in using the WWW is 95%. The majority of users (99.3%) report checking their e-mail at least once a day. Most respondents report using Internet services from home as well as other locations (73%) and for personal reasons (65%). Users access the Internet most frequently during the evening hours, from 5 P.M. until midnight. These are consistent with national data showing more home

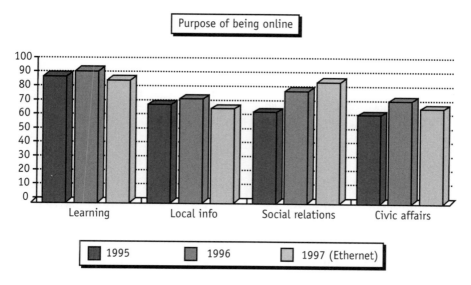

Figure 5.1 Helpfulness of Internet for different purposes.

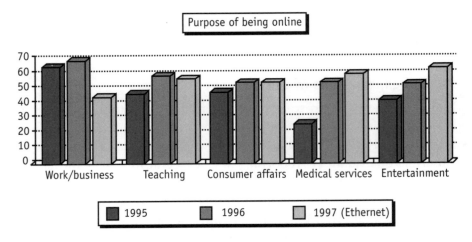

Figure 5.2   Helpfulness of Internet for different purposes.

Table 5.1
Helpfulness of Internet for Different Purposes: Early Subscribers

Purpose	1995	1996	1997 (Ethernet)
Learning	90%	94%	88%
Local info	70	74	67
Social relations	64	79	86
Civic affairs	62	72	68
Work/business	65	69	46
Teaching	47	59	57
Consumer affairs	48	54	54
Medical services	26	54	60
Entertainment	42	53	64

use of Internet than use from the workplace. The majority (77%) of users reports accessing the Internet in the last 24 hours; respondents were connected an average of about an hour in their last Internet session.

In considering longitudinal use of the BEV home page, users report that the links they access most frequently are local or regional sites, specifically: (1) "library and reference tools" and (2) "regional information" (tie), (3) a health care center, (4) Virginia Tech Web, and (5) Virginia resources. Eighty-three percent use a modem to connect to the Internet; 3% use off-campus ISDN; 11% use off-campus Ethernet; 30% use on-campus Ethernet; 7% use on-campus CBX (percentages total to more than 100% because people access the Internet from multiple locations).

Trends among the general population of Blacksburg

Beginning in 1994, BEV and the Town of Blacksburg enclosed an Internet survey in 4,000 *About Town* newsletters annually. The surveys went to randomly selected households within each census block of town, with returns of 332, 334, 224, 202, and 172, respectively. While these response rates are low, they also reflect a more conservative estimate of technology use. The large student population in the off-campus apartments is not included in the sample because they do not directly pay a water bill and therefore do not receive the town newsletter. The students are typically Internet users, particularly since August 1998 when the university required all students to own a computer at the start of the academic year. Therefore, the reported results are conservative estimates of trends in Internet behavior and effects among the general population of the town.

The profile of the average respondent at the beginning of 1999 (January survey round) is similar to that of previous years: Roughly half of respondents are female (55%); the average age is 43 years; and 42% of all respondents have completed graduate school. Most respondents (59%) are members of a church or local club. The average length of residence in Blacksburg is 14 years.

Table 5.2 shows trends in computer ownership, use of e-mail (proxy for Internet access), and computer and network experience—that is, the percentage of respondents reporting they are "somewhat experienced" or "very experienced," the two highest points on a four point scale.

The use of e-mail is the most popular service in Blacksburg area, which is consistent with findings in national studies. Respondents use e-mail primarily with friends and family, followed by coworkers and other groups. Table 5.3 lists those with whom (in percentages) the users

Table 5.2
Trends in Computer and Network Use

	1994	1995	1996	1997	1999
Home computer	74%	75%	79%	81%	81%
Use e-mail/Internet	—	62	69	81	86
Computer experience	75	72	71	80	76
Network experience	38	40	51	67	72

Table 5.3
E-Mail Correspondence With Whom?

	1995	1996	1997	1999
Friends non-Blacksburg	61%	55%	79%	84%
Friends Blacksburg	50	49	70	65
Family	49	52	76	66
National interest groups	49	44	46	54
Coworkers	43	46	61	60
Local interest groups	42	46	41	30
Teachers	37	38	53	46
Classmates	32	34	52	42
Neighbors	30	37	29	24
Support group	25	29	31	28
Church members	21	27	26	26

in the town of Blacksburg expect the BEV to be somewhat or very helpful to communicate. Figure 5.3 (a–b) charts these results.

The e-mail correspondence among non-Blacksburg residents of Montgomery County differs somewhat from that of Blacksburg residents. As with Blacksburg residents, the majority of non-Blacksburg residents use e-mail to correspond with family (65%) and friends outside the area (71%). However, a higher percentage of non-Blacksburg residents report using e-mail with national groups, support groups, and club or church members. See Table 5.4 for these figures.

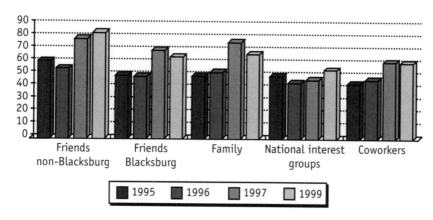

Figure 5.3(a)   E-mail correspondence.

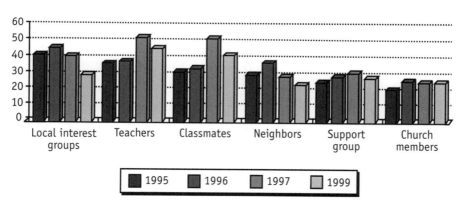

Figure 5.3(b)   E-mail correspondence.

The majority (88% in 1998) of Blacksburg residents report that they expect the Internet to be "somewhat helpful" or "very helpful" for accessing general information. Forty percent report that it will be somewhat or very helpful for civic affairs (1999). A majority also expects the Internet to be "somewhat" or "very helpful" for accessing local news and information. The large increase in the use of commercial services reflects the general trend of commercialization of the Internet, with many more choices by 1999 for news and information from for-profit sites. Table 5.5 shows trends among users and nonusers in the general Blacksburg population who expect the Internet to be helpful for various purposes. Figure 5.4 charts these results.

Table 5.4
E-mail With Whom? Blacksburg vs. the County 1998–99

	Blacksburg	County
Friends outside the area	84%	71%
Family	66	65
Friends in the area	65	49
National interest groups	54	55
Coworkers	60	52
Local interest groups	30	29
Teachers	46	40
Classmates	42	39
Neighbors	24	21
Support group	28	34
Club or church members	26	36

Table 5.5
Helpfulness of Internet for Different Purposes; Blacksburg

	1995	1996	1997	1999
Local news/information	63%	55%	61%	50%
Health/safety information	40	40	40	48
Commercial services	29	29	32	57
Playing online games	13	18	15	15

Blacksburg residents and non-Blacksburg residents of Montgomery County disagree on the helpfulness of the Internet for various purposes. The majority of both residential groups reports that the Internet is "somewhat helpful" or "very helpful" (the top two scores on a four-point scale) for general information. For other purposes, however, the percentages differ widely. The majority of non-Blacksburg residents (66%) report that the Internet is "somewhat helpful" or "very helpful" for health and safety information, while only a minority (33%) expects it to be helpful for civic affairs. Table 5.6 lists statistics on the helpfulness of the Internet for various purposes in Blacksburg and Montgomery County in 1998–99.

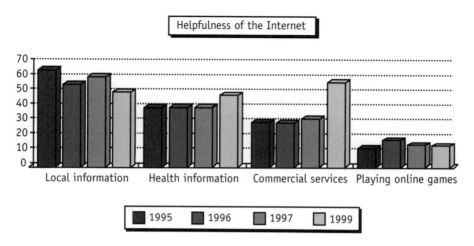

Figure 5.4 Helpfulness of the Internet for various purposes.

Table 5.6

Helpfulness of Internet for Different Purposes: Blacksburg vs. the County 1998–99

	Blacksburg	County
General information	88%	77%
Civic affairs	40	33
Local news and info	50	47
Health and safety info	48	66
Commercial services	57	38
Online games	15	46

The impact of networking on community

Computer networks such as the Internet allow interaction among groups of people. As such, they have the potential in geographic communities of supporting and extending social relationships. The quality of life in a community with dense social networks, high levels of trust, and norms of mutual reciprocity, what Robert Putnam (1995) calls its "social capital," is higher than the quality of life in communities with low social capital.

The concentration of local information and services, local discussion groups, and other locally focused material on the BEV Web pages provide the environment in which social networks, social trust, and norms of mutual reciprocity can be supported, or possibly even enhanced. From the outset of the project, users expressed their interest and satisfaction in being able to be more connected to their community. As noted in Table 5.1 regarding social relations, 79% of respondents expected the Internet (BEV) to be somewhat helpful or very helpful (two highest points on a four-point scale), up from 64% two years earlier.

In a random sample telephone survey conducted in November 1996 of the local calling area (Blacksburg, Montgomery County, and environs), results showed that outside Blacksburg, there was a statistically significant correlation between community involvement and access to the Internet ($p < 0.05$). That is, Internet users are significantly higher on measures of community involvement than nonusers. Among Blacksburg residents, there is a positive but not significant correlation between Internet access and community involvement. This could be a ceiling effect, since the town already has dense social networks and high levels of community involvement. Among the total population of respondents (Blacksburg, Montgomery County, and environs), about one-fifth of the Internet users (22%) report that they are "more involved in their local community since getting on the Internet." This proportion of the population scores significantly higher on measures of community involvement than those reporting that they are "equally involved" or "less involved in their local community since getting on the Internet." This would imply that the Internet has increased the involvement of people who were somewhat involved already, but whose level of activity has declined over time, as Putnam has enumerated.

Local business trends 1995–97

Among Blacksburg businesses using the Internet and responding in two survey rounds (1995 and 1997), most (65%) are small with fewer than 10 employees. There is a variety of businesses represented, although most are in the services sector offering retail, consulting, or professional services. Most (61% in both rounds) have five or more years experience with computing; about a third (30% in 1995, 39% in 1997) have more than five computers in their organization. See Table 5.7 for the complete results and Figure 5.5 for a chart of the results.

Over a third of the businesses (37% in both rounds) use the computer between 26 and 50 hours a week. The majority of respondents (86% in 1995 and 84% in 1997) use Windows. Most (80% in 1995 and 70% in 1997) report having one or two computers connected to the Internet. They use the networked computers up to 20 hours a week (90% in 1995, 81% in 1997). The majority (75% in 1995, 67% in 1997) has dialup access to the Internet. The drop in dialup access is due to businesses shifting to direct connections via Ethernet (10-Mbps) or T-1 lines (1.45-Mbps).

Table 5.7
Business Size

Number of Employees	1995	1997
Fewer than 10	65%	65%
10–50 employees	13%	26%
50–100 employees	13%	5%
More than 100	9%	4%

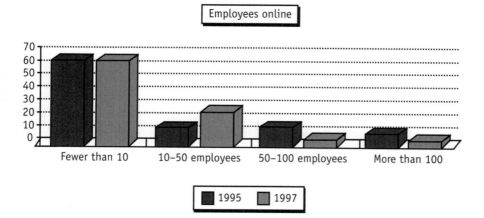

Figure 5.5   Business size (chart).

Business users' motivation and impact

Among the reasons noted for using the Internet in the earlier survey round, many businesses (65%) reported wanting to take advantage of the free advertising offered by a "business card" listing on the BEV Village Mall, however, that service was discontinued because there was no need to provide it later on in the project. Moreover, many respondents (70%) expressed interest in the technology for its own sake. In 1997, many respondents (63%) reported they are still interested in the free "business card" listing, but respondents interested in the technology "for its own sake" dropped from 70% to 52%. As for increasing contacts via the Internet, respondents seeking to increase client contacts drops slightly (from 84% in 1995 to 78% in 1997), while the percentage seeking to increase contact with suppliers rises slightly (from 8% in 1995 to 11% in 1997). It should be noted that a business putting a listing online merely because it is "free" is probably not going to see much increase in business or gain much value from the limited exposure free listings offer. Businesses that have been successful online have achieved that by integrating the new technology into their strategic marketing and planning efforts. Table 5.8 lists businesses' reasons for listing online.

In the earlier survey round, only 14% of respondents reported an increase (up to 10%) in contacts per month (with suppliers and clients), and 4% reported increases from 11–20%, resulting from the Internet. By 1997, 27% of respondents noted up to 10% increases in contacts per month, and one company, a Web distributor of books, noted increases between 41 and 50% in contacts with clients and suppliers. Table 5.9 lists respondents' increase in contacts per month in 1995 and 1997.

Table 5.8
Reasons for Listing Online

	1995	1997
It is free	65%	63%
Interest in technology	70%	52%
To increase contacts with clients	84%	78%
To increase contacts with suppliers	8%	13%

Table 5.9
Increase in Contacts (per month)

	1995	1997
None	82%	70%
1–10% increase	14%	27%
11–20% increase	4%	0%
31–40% increase	0%	0%
41–50% increase	0%	3%

On the revenue side, only 5% of respondents in 1995 noted increases (up to 10%); by 1997, 17% of respondents noted up to 10% increases in revenue generation per month resulting from their Internet site. The Web distribution business noted revenue increases between 41 and 50%. Table 5.10 lists respondents' revenue increases in 1995 and 1997.

Online commercial transactions

While online commercial transactions are not currently offered as a service to local businesses (as of 1998), the BEV is interested in making it possible for businesses to conduct such transactions in the near future. A major disincentive to online transactions is the lack of a standard in that area and the fact that existing transaction systems are very expensive. A recently introduced product (1999) required a $50,000 "training" fee

Table 5.10
Increase in Revenue (per month)

	1995	1997
None	95%	80%
1–10% increase	5%	17%
11–20% increase	0%	1%
21–30% increase	0%	0%
31–40% increase	0%	0%
41–50% increase	0%	2%

up front and wanted a 30% commission on every transaction. Not only is this out of reach of most businesses in the world, the transactions fees are just plain silly. If the partially automated credit card systems we use today can be profitable with transaction fees of 3–5%, a fully automated, Internet-based system should be less, starting at 1% and going down from there. If "pay by the drink" sales of information (i.e., paying a small amount, like one cent, to be able to read the daily news from a news site) is going to develop, transaction systems must reflect the realities of the marketplace.

There is evidence of a trend away from interest in online commercial transactions, dropping from 60% in 1995 to 53% in 1997. Among respondents not interested in conducting online sales, one-third of respondents in both rounds (32% in 1995, 31% in 1997) noted that online commercial transactions were not applicable to their business. Only a few companies reported concern about data security (8% in 1995, 11% in 1997).

Local businesses not using the Internet

Among the 92 respondents *not* using the Internet that were surveyed in the 1995 round, most (72%) indicated that they do not list information about their business on the Internet (the BEV Village Mall Web pages) because they do not use computers in their business, or, if they do use computers, they do not have a network connection (28%). Other reasons they could have noted in 1995 (listed below) were not a consideration. Table 5.11 lists the reasons businesses did not list online in 1995 and 1997; these results are depicted graphically in Figure 5.6.

Table 5.11
Reasons for Not Listing Online

	1995	1997
Don't use computers	72%	6%
Not networked	28%	22%
Cost	0%	14%
No market application	0%	28%
Other	0%	24%

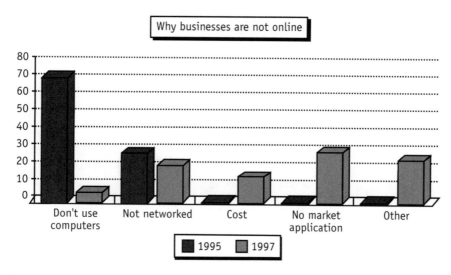

Figure 5.6   Reasons for not listing online.

Table 5.11 shows that in the 1997 survey round, only 6% of respondents noted that they do not use computers in their business; 22% said that they do not have a network connection. Other factors noted as reasons why businesses do not list themselves on the BEV site included "no market application," "cost," and "other."

In 1995, almost all respondents (95%) noted that technical training and education were important motivating factors for using the Internet in the future. In 1997, 94% of respondents noted that one to five of their employees had taken Internet-related training classes in the previous year; this compares favorably with 95% for this variable among businesses that *do* use the Internet. In the 1997 round, training and education were motivating factors for only 29% of businesses *not* using the Internet. Table 5.12 lists other important motivating factors for using the Internet in the future (noted by 1997 respondents but not noted by any of the 1995 respondents).

Public access and training at the public library

Since the beginning of the BEV project in 1993, the Montgomery-Floyd Regional Library (MFRL) has promoted an ongoing program to provide

Table 5.12

Incentives for Future Use

	1995	1997
Training	94%	29%
Reduced cost	0%	22%
Large market base	0%	25%
Revenue potential	0%	51%
Decreased cost for marketing	0%	22%
Nothing	6%	10%

universal information access for the residents of the diverse communities of the New River Valley. Under the Telecommunications and Information Infrastructure Assistance Program (TIIAP) grant, MFRL along with its partners, the BEV and Montgomery County Schools, has proactively and aggressively addressed the issue of disparity in information access. MFRL has done this by providing not only easy access to information resources on the Internet but also the training required to make effective use of these resources. This section summarizes some of the developments leading up to service and training through the library, as well as survey findings from patrons using the facilities. MFRL began its first program of Internet training for staff in October 1993 at the Blacksburg Area Branch (BAB). Public Internet access officially opened in January 1994 with public training beginning shortly thereafter at this site. The MFRL computer specialist and the electronic reference librarian conducted these classes and demonstrations.[2]

Initially, a single T-1 data line fed seven computers at the BAB. MFRL expanded this public Internet access to include the other two MFRL libraries[3] via a frame relay WAN integrated with SIRSI, the MFRL automation system.[4] T-1 lines were installed between Blacksburg and Christiansburg, providing a 1.55-Mbps data transmission and adequate

---

2. Referred to as BEV librarian in other MFRL reports on LCSA-funded activities.

3. Headquarters Library in the Town of Christiansburg and the Jessie Peterman Branch Library in the Town of Floyd.

bandwidth to facilitate use of advanced applications. The Floyd library was connected to the WAN via a 56-Kbps data line due to the site's smaller size, fewer computers, and limited budget. Providing Internet access at all three branches of the MFRL necessitated providing additional training to both staff and to the public. Previous MFRL studies conducted under Library Services and Construction Act (LSCA) grants revealed that familiarizing existing staff with Internet navigation protocols required an initial training session. Also determined was the need for three follow-up sessions and that even after these training sessions, some staff did not feel comfortable with answering patron questions about the Internet or its usage.[5] The library believed that it was necessary to hire personnel specific to the duties of public Internet training. The goal was to help the public achieve an appreciable level of network literacy to help eliminate the fragmentation of local communities into segments of information "haves" and information "have-nots."

In 1995, MFRL, the Montgomery County Public Schools, and the BEV submitted a joint application for federal assistance to the National Telecommunications and Information Administration (NTIA) through TIIAP. The grant was approved in the fall of 1995. As a result, a full-time Internet trainer was hired at the library. The hiring of the trainer was delayed several months in order to set up the logistics of paying the NTIA-funded trainer through the Montgomery County payroll.

In April 1996, training began at all branches of the MFRL system: Blacksburg, Christiansburg and Floyd. This project was the first time MFRL tried to implement Internet training at all three branches. For 18 months, this grant-funded MFRL Internet trainer maintained a training schedule that rotated between branches. This training program covered demand for both public and staff training in three communities of widely varied composition and culture.

The purpose of the MFRL surveys was to compare the basic demographics of trainees with those from previous surveys conducted at the

---

4. SIRSI Corporation's Unicorn Collection Management System is now in use at all MFRL branches.

5. "Though a few staff members still remain uncomfortable with using the BEV system, this small minority may be unreachable for a variety of individual and organizational reasons," according to Bradley Nash, Jr., Perceptions of Internet Training staff member.

BAB of the MFRL. Another purpose was to use the survey responses to help evaluate the scope and effectiveness of the MFRL training for public access to the Internet during the time period of May 1996 through May 1997.

Findings

The vast majority of trainees fell into the "adult" category. The average age of participants system-wide for the first 227 surveys was determined to be 47, as opposed to 38 and 37, as calculated from RSS I (1993) and RSS II (1995), respectively.[6] So, the addition of the demographics of the Town of Christiansburg and Floyd County skewed the results toward higher age brackets. Nineteen percent of the trainees were in the 65-years-plus bracket, showing a real demand amongst the senior community for training in technology. In fact, the Internet trainer specifically noted the enthusiasm and commitment of the seniors who attended her classes.[7]

As in the RSS I and RSS II, gender distribution of recent trainees was found to strongly favor "female." This tends to indicate that the women of the area have as much interest in the Internet as the men, and will use it if the facilities are available.

Although 18% of respondents to the current survey reported a high school degree or less formal education, as opposed to 12% and 10% in the Blacksburg-based surveys to which we are comparing data, this still leaves 81% of current respondents reporting at least some college schooling. Therefore, in spite of the rural atmosphere of the area, the overall level of education is quite high.

---

6. It should be noted that the respondents to the Regional System Surveys (RSS) were simply library users, not specifically trainees.

7. "The only other element I'd like to point out is the number, and enthusiasm of, the senior citizens who were reached during my tenure. They were by far my best students as a group and they usually came from the furthest—from no computer skills to PC owners who regularly utilize both e-mail and the Web for both pleasure (predominantly) and their financial interests (i.e., tracing stocks and finding part-time jobs)," Kimberly V. Evans in "Notes from the Ex-Internet Trainer," letter to the network services clerk/typists, August 1997.

Regarding the scope and effectiveness of the MFRL public Internet training for the period of May 1996 through May 1997, it should be pointed out that at least 1,002 people were trained during this time period. Immediately following the beginning of WWW basics classes, popularity of the Internet access terminals at the MFRL branches jumped tremendously. Suddenly, the library didn't have enough terminals to meet demand. It was not uncommon for patrons to encounter a wait time of 15–60 minutes on an average afternoon for a chance to use the Internet. MFRL responded to this situation by adding more workstations, and as the training proceeded over the course of the year, demand continued to increase. Librarians noticed many patrons returning for regular Internet sessions. The result at the Christiansburg Branch was that two workstations were designated for SIRSI online catalog access only, while all seven additional workstations provided Internet access.

In terms of increasing public awareness of the Internet and popularity of its use in all three MFRL branches, the training program was a success. The Internet trainer reported that, "We reached many people who would otherwise never have had exposure to either the Internet or computers in general, in a systematic way that both increased their skills and awareness."[8] The Internet trainer related other stories of success, including the experience of several first-year French students at Floyd High School. After taking the WWW basics class, they were able to contact students in France and become virtual penpals with these students.[9] The trainer also related that "[A y]oung couple attended several of my classes including HTML basics. They had recently had their first child…They went on to create a Web page with shots of their baby's pictures to keep their family…up to date with the baby development."[10] Another observation relevant to the success of the project: "An unemployed mother without computer skills attended all four of my courses and through the monster board Web site found a job paying quite well enabling her to work at home and continue to care for her young children."[11]

---

8. Ibid.

9. Ibid.

10. Ibid.

11. Ibid.

Selected Bibliography

Kavanaugh, Andrea, "Highlights of Internet Users: 1994–99," *About Town,* Blacksburg, VA: Town of Blacksburg, 1999.

Kavanaugh, Andrea, and Scott Patterson, "The Impact of Computer Networking on Social Capital," paper presented at the National Communication Association, November 1998.

Kohut, Andrew, "The Internet News Audience Goes Ordinary," Washington, D.C.: The Pew Research Center, 1999.

Patterson, Scott, and Andrea Kavanaugh, "Building Critical Mass in Community Networks," paper presented at the Speech Communication Association, August 1996.

Putnam, Robert, "Bowling Alone: America's Declining Social Capital." *Journal of Democracy.* 1995, 6:67–78.

Turow, Joseph, "The Internet and the Family," Washington, D.C.: The Annenberg Public Policy Center, 1999.

6

Community Dynamics and the BEV Senior Citizens Group

by Federico Casalegno, Sorbonne University

THE AIM OF MY WORK was to try to understand the dynamics of a community and to show new forms of networked communications mediated by computer. To attain this goal and to better understand a network community, I have set up a general scheme that considers five major values:

- The domus;
- The virtureal space;
- The communication;
- The genius loci;
- The leader.

The scheme is a work in progress and, like all hypotheses, must be verified. In this chapter, I have used the scheme to analyze the interviews that I carried out in Blacksburg in May 1996. I questioned 14 members of BEV Seniors group to understand the internal dynamics of the group and to test the grid that I am developing. With this aim, I will explain the values that characterize the interpretation grid, and I will use it to understand the BEV Seniors group. (The basic questionnaire used during my talks is provided in the appendix at the end of the chapter).

The domus of the BEV Seniors group

The grid

To understand the value of domus in a network community, consider the following.

1. The physical domiciliation of the network (city, neighborhood, public/private place, etc.) and the cultural environment (social relationship, history of the community, town or region, etc.);

2. The goals, which can be categorized as economic, entertainment, social, or other, and which can be declared and/or undeclared;

3. The membership: How we became members of a particular community, the advantages of membership, the obligations of membership, and the rules and the behaviors of the members of the community. (Figure 6.1 displays an interpretation grid for network communities.)

The BEV Seniors group

The first observation to make is that one becomes a member and adheres to the group by direct knowledge of someone else who is already part of the group. There is not a true active recruitment policy for new members, but the beginners are aware of the activities of the group thanks to a friend or relative. The "leaders" speak about the group while trying to publicize their activities—without a formal program of recruitment. Sometimes, in fact, a son or other family member informs a senior about the existence of the BEV Seniors group and encourages him or her to join.

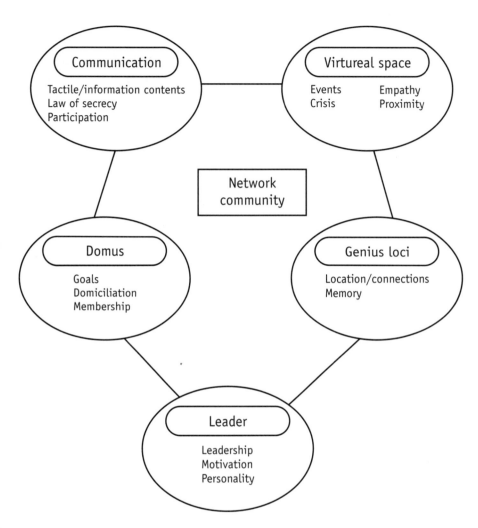

Figure 6.1 Interpretation grid for network communities.

The informing family member or friend is then a true agent of socialization; in some cases, such agents do not only speak to seniors about the group but even buy (and/or install) the computer so that the senior can became a member of the group.

The fact that the press speaks about the group also makes people increasingly aware of the BEV Seniors. Furthermore, since retired people typically belong to several groups, people can learn about the existence of the BEV Seniors group through membership in other associations or organizations. The merging of the activities of different associations allows an exchange of information between various groups.

There is an important difference between the BEV Seniors and the Blacksburg Senior Group: The first is a group of senior citizens that wants to learn how to use new technologies, while the second is an association for all senior citizens of the town. Participants in the latter group are not interested specifically in new technologies; they mainly arrange social activities such as dinners, trips, bridge tournaments, and educational activities such as aerobics, art classes, and conferences.

The seniors who I questioned are affiliated with several associations at the same time (e.g., New Dimensions, for the retired from Virginia Tech University; various religious associations; or the American Association of Retired People (AARP)). The declared purpose of the BEV Seniors is to make it possible for senior citizens to learn how to use computers and, particularly, the Internet.

The BEV Seniors do not consider there to be a single goal of the group, apart from training in new technologies of communication. The BEV Seniors group is an abstract group whose members can draw from it whatever interests them. In order to explain this principle, one of the interviewees adapted a famous remark of John F. Kennedy's, saying "BEV Seniors is whatever you want it to be."

The primary interest of the members is to learn how to use the computer to interact and communicate with others—using it as an instrument and a resource to solve the practical problems of everyday life. A benefit for some members, in addition to learning how to use computers, is that the group meets monthly, without structuring purposes into the meetings. As some who were interviewed said:

"[BEV Seniors] is not like the Rotary or some great social organization that meets to do projects."

"The BEV Seniors group is a local organization that wishes to help the local population of retired persons of Blacksburg. Its goal is not to improve the life of retired in general but to better integrate the local senior citizens among themselves and with their social environment."

"The purpose of the BEV Seniors is not really how to improve seniors in general. Of the 75,000 people who are in this area, there [are] perhaps 10,000 seniors, 8,000 seniors, I don't know. So the purpose of the BEV Seniors is not to improve the status of senior citizens in general; we just deal with our small group."

To become member a of the group, there are two qualifications; it is necessary to live in Montgomery County and to be 55 years of age or older. There is also a kind of ethical code that all members must be helped—especially with computers. One does not pay dues, but as an initiatory rite (a kind of hazing), it is necessary to send an e-mail to the BEV Seniors newsgroup and to introduce oneself and list one's interests, background, passions, etc.

It should be stressed, however, that this group is organizing itself more and more. With approximately one new member per week, BEV Seniors had grown to 134 members in 1996. The steering committee, created in January 1997, meets twice monthly in order to decide what actions to take and to direct the group. Within this framework, the group has prepared a handbook to learn how to surf on the Internet and has searched for used computers to give courses to members. In the fall of 1998, the BEV Seniors opened its own computer lab in the new Senior Center, a recent addition to the Town Recreation Center. Prior to this time, the group had relied primarily on computer lab spaced provided by Virginia Tech's New Media Center.

Overall, the group is displaying a tendency to give more and more rules to itself and to organize itself more optimally. It is necessary to follow this evolution to see whether this growing organization increases the imposition of rules or obligations.

What are the benefits of belonging to the BEV Seniors group? First of all, it allows members to share experiences and fragments of life. You can learn with the others to use the computer, and you can feel truly "interconnected". Also, one can follow and take part in the activities of the group. (I will reconsider this aspect later). Second, learning how to use computers is helpful in many ways. Mainly, it is important to remember that it can be frustrating for an inhabitant of Blacksburg not to know how to use the Internet; to live in what is regarded as "the most wired town" in the world (according to *USA Today*) and not know how to use a computer seems paradoxical and could give one a feeling of exclusion. For an older person, moreover, it can mean remaining completely apart from the active life of the town. This point was well supported by one interviewee:

"…well I think people who do not touch the computer must feel very left out in today's world, and using the computer makes me, well … I am included in a lot of information and knowledge and conversation that I wouldn't be if I didn't use the computer. Now being a member, a senior member of BEV net is a part of using a computer and I guess that gives you a certain sense of belonging, not to the BEV net but belonging to the universe."

Feeling part of the "new world," or having information and knowing what occurs in our environment, tightens our links within our community. New BEV Seniors participants learning how to use the computer can benefit from the experience of more experienced members of the group. Volunteers regularly show members how to connect to the Internet and how to use e-mail, the Web, the newsgroups, and so on. These demonstrations also provide members with an opportunity for contact.

"There's somebody out there whom maybe I don't even know who is in the same kind of age category I am and maybe in the same category of ignorance—if I can use the term—about what to do with this thing, and I can talk to him on e-mail and tell him what my problem is and find out if they [sic] know and one fellow will say, 'well, I don't know, ask so and so.' So, then I'll e-mail that person, or maybe that person will have seen it on his screen and will come in and say 'I've had this experience before. Let me tell you what I did.'"

Formal BEV Seniors meetings take place in a room of the town's recreation center. The steering committee organizes lectures on many subjects of interest, while providing an opportunity for the members to meet face-to-face.

The virtureal space of the BEV Seniors group
The grid
To understand the value of virtureal space (the synergy between virtual and real space) in a network community, consider the following.

1. The events that may be proposed from the bottom or imposed from the top of the community and can be organized for the community itself (i.e., dinners, parties for community members, conferences, games, and trips), for the larger community (e.g., clean-up days and garden tours), or for other communities (such as helping young/old people and collecting food and clothes for poor countries);

2. A crisis that could change the dynamics of and affect the stability of the community;

3. Empathy, or the moments that make people stay together (online and in real life) and feel closer;

4. The proximity (physical and cultural): People share spaces, hobbies, and a territory (with a real dimension but also a symbolic one).

The BEV Seniors group
To date, there has never been a genuine crisis in the BEV Seniors group. (There was, nonetheless, a time of dissatisfaction when Virginia Tech could no longer give modem access to members who were not affiliated with the university. BEV seniors not affiliated with Virginia Tech were obliged to obtain Internet access from any one of several new private providers, at a cost of about $16–20 dollars per month.)

When I questioned the members of BEV Seniors about whether they've experienced moments of crisis in the group, their answer was negative but nevertheless reflected that their group is not very organized or structured. We can therefore conclude that at the present time there are no power struggles or fights for prestige or money in the group and thus there are no reasons for conflicts between members. However, the growing numbers of seniors involved in the organization and its decision-making structure (the steering committee, which allocates scarce resources to find computers and which has produced a users' handbook) may cause problems in the future.

Currently, there are three BEV Seniors projects that represent the primary occupation of the group and, at the same time, are representative of the group's empathy:

1. The first project concerns certain members of the group who are seeking computers for installation in a room of the town hall or recreation center to give courses to the other members of the group.

2. The Nostalgia project, which aims to collect all stories relating to Blacksburg, a town whose bicentennial celebration occurred in 1998. The goal of this project is to use the memories of the senior citizens to build on the Web an image of the city such as it was in time past: how it has changed; which stores have always been active; which stores have disappeared; how the town planning of the city has changed; and a description of the customs, manners, or practices of former residents. Anyone can write a part of this story and post it directly on the Web. People can also send photographs of different places, houses, or stores; comment on the stories of others; and add Web links to other sites that could interest visitors to the Nostalgia project Web site.

3. The last project is the youth-senior mailing list. This endeavor is about sharing experiences and exchanging points of view between youth (a class of 12-year-old students) and BEV Seniors. Thanks to e-mail, the young people can pose questions to seniors such as the following: What did you make for Christmas? What did you eat? What were your best toys? What was school like in your day? In short, the young students question seniors in order to learn about former everyday life in a personal way. Concurrent to the youth-senior online activity, participants have organized two face-to-face meetings: The first occurred when the students invited seniors to their school; the second occurred when the seniors invited the young people for a picnic at the town park. In addition to these exchanges, the BEV Seniors organize meetings

in a room in the town recreation center, where they can speak about the actions of the group and plan activities to come. During my talks with them, the BEV Seniors indicated that they liked these meetings very much. The most friendly moments were right before and after the meetings, as they could speak freely between themselves.

"I think the meetings make the group feel closer because you see each other in person (and that would be both the steering committee meetings and the other, the general meetings). Usually there is a little social activity before and a little bit after. The social activity is coffee, say[ing] hello; sometimes it's computer talk, and sometimes it's something else."

It seems that the BEV Seniors like to socialize and to discuss things, but within the framework of an organized meeting. On the other hand, they prefer not to have too many formal meetings.

It is also important to note that the exchanges on the listserv, in synergy with the meetings, makes it possible for members "to give a name to a known face." Many of them knew each other by sight before joining the BEV Seniors, but it is thanks to their participation in the listserv and the meetings that they know each other better. Now when they cross each other downtown they stop to exchange a few words, something they did not do before.

Thus, even if they do not directly organize meetings between members of the group, being part of the BEV Seniors group allows members to be discovered and to meet others. From this angle we see a homeopathic integration of the seniors in the community. It is a fundamental solution to many of the problems related to exclusion in the senior stage of life.

"Well, the thing is unless I happen to [have] actually physically know[n] someone in the past, it's unlikely that I would have talked over the problem, whatever. Now, there is a focal point that we can all come to and exchange ideas and discuss things about them, which I wouldn't have known [about], for example, that D. W. was interested in this, that, or the other. I would not have known that J. S. was really into this genealogy ... it just happened, happened by chance to come up in a conversation socially. So, I think it's [facilitated] a dramatic increase in our interactions."

Sometimes the exchanges in the listserv make it possible for participants to discover that there are other members who share their interests. Thus, the listserv can allow BEV Seniors members to begin a direct relationship without going through organized, face-to-face meetings.

The communication of the BEV Seniors group

The grid

To understand the value of communication in a network community, consider the following.

1. The difference between the tactile information (poor content and rich emotional exchange) and information based on the contents (high information exchange).

2. The participation: The more people are involved in sending information (online or in real life) the more they reinforce the feeling of belonging to the community.

3. The laws of secrecy (information, behavior, rites, codes, etc.) that establish contact with others within the confines of a limited group, while conditioning the attitude of the group toward external forces.

The BEV Seniors group

In the grid of interpretation, I propose that for each community it is important to underline the laws of secrecy, or the mystery that connects the initiates. If one takes the example of the presentation on the "virtual reality cave" (explained below), one realizes how the meetings between seniors have the double function of tightening the links between members of a community and members' links with the greater community (e.g., the town and the district).

Faculty at Virginia Tech have established a virtual environment (the virtual reality cave at http://www.cave.vt.edu) that allows researchers to have the impression of being inside the studied object. (For example, doctors have the visual simulation of being inside the human body.) A demonstration of virtual reality to the BEV Seniors group permitted them to have a specific knowledge of a particular object; not many people among the seniors of the community as a whole or even all the inhabitants of Blacksburg have had such an experience. The virtual reality cave demonstration not only allowed members to share a secret but it also made it possible for BEV Seniors to feel nearer to the "active society" and to the avant garde of technological research. The fact that the members belong to the same age group enables them to better understand each other.

When a BEV Seniors member must go to another member's house to teach him or her how to use the computer, he or she truly has the impression of belonging to the same group as that person—and therefore having the same problems and concerns. BEV Seniors can more easily enter into contact with other members and be put at ease. Sending a small portrait of oneself to the listserv makes it possible for members to know each other better, and it plays the initiatory rite of strengthening community links.

The subjects of conversation, finally, constitute the true secrecy that levels and links the group. Participation in the BEV Seniors' listserv is reserved exclusively for the BEV Seniors group: They share the contents of the messages in this network only with themselves. Here, there are no predetermined subjects, and any member can send messages. This type of exchange belongs to the tactical communication category, because it helps to reinforce link between members.

> "Oh, I expect that one of the things that happens with the BEV Seniors is that they communicate with each other on this. I mean there is a list, and people want to send the message to all the BEV Seniors. And you get a lot of useless messages, but it's probably still kind of fun. It is. They pass information back and forth about what they're doing or what the weather is going to be or who they think is going to win a football game or this kind of thing."

How do BEV Seniors use the system of communication? The ultimate purpose is to use the network and the computers as resources for practical everyday needs. Some of them use computer networking to obtain the menu of the restaurant, check specials, print out a coupon, or make a reservation. Some organize trips, reserve tickets, and find the best travel plans, hotels, and tourist information. Others manage their business and financial affairs on the computer.

Since users can do so much from their house by computer, I asked several questions about whether they reduce face-to-face contact with people and "the real world." The questions were inspired by the ideals of the technoluddist. The answer is simple and is summed up by the response of one interviewee: "Since I can manage my bank account by computer, I can use my spare time for other, much more interesting, activities."

One devotee of art found a site with information about artists, artistic periods, etc. that one could download, print, and frame for his or her own home. These examples show how the members of BEV Seniors use the network to deepen their own interests and to seek practical information to help them with everyday life. The most used function, however, is not the Web but e-mail and the listserv.

> "I find that one of the things that's kind of interesting are exchanges on the listserv. For instance, [one] member wrote in and asked if anybody had patterns for dolls...I knew I had some older—about 10–15-year-old—magazines with some crafts and art work in [them]. So, anyway, I looked [them] up … and wrote back. …you can exchange things like that, that don't have anything to do with computers at all, but have to do with interest[s] in life. And I am not one that has happened to chime in on it, but I know a lot of people that have had conversations about gardening and things like that, and so they maybe exchange ideas or exchange bragging rights about the first tomato that's ripe or something like that … I think that … there is sort of an obligation in the sense that if you want to get back some things that are valuable to your interest, you need to occasionally contribute something that is interesting to you."

Many BEV seniors have grandchildren who study or work in another part of the country. E-mail quickly becomes the preferred communication system between BEV Seniors members and relatives for its low cost and the numerous benefits of asynchronous communication. With e-mail, BEV Seniors can communicate their opinions, joys, fears, and hopes to a considerable number of users with little effort (only one e-mail message). Phone calls to join the 134 people of the BEV Seniors, on the other hand, would be time-consuming and exhausting!

For the majority of members, the listserv is used for tactile communication—that is, to remain in touch with close friends who are geographically distant. For most, the computer network enables them to remain in contact with others and, metaphorically, to touch their lives. A member of the group, who writes to her 25 women friends regularly, says she appreciated very much a recent birthday gift. Her sister, who lives across the country, used the BEV Web site to order pizza and have it delivered to the interviewee's home for her birthday. When she opened the door, the delivery man sang "Happy Birthday." She says that the Internet really brings people closer.

The data related to BEV Seniors participation are surprising. Members do not do the bulk of the work on the Web site of the group. It is the steering committee that takes care of it. Moreover, members of the BEV Seniors do not connect to their home page frequently. (Only one among the people interviewed said that he/she looks at it sometimes.) The majority of them say that they read all the messages of the listserv and check their e-mail at least twice a day (sometimes three times a day). Only one person said that he sends messages to the listserv regularly. One of the leaders, on the other hand, said that he sent up to four messages per week, especially as the BEV Seniors group was establishing itself. If he did not have any specific news, he sent a simple greeting to the members and wished them a good day. It is in this context that I formulated the assumption that the members of the group can be defined as *voyeurs actifs:* They do not miss any message nor movement in the listserv, but their activities are limited to observation.

> "I am not interested in sports … But it is nice to see that there are these relationships."

> "I am a reluctant participant because I feel that I am so inadequate with what I know. I enjoy reading all the things the other members write, but I seldom write something myself."

These remarks show that the listserv is a window on the community and ensures a link among members. The listserv nourishes the seniors and sustains their interest in their community.

The Web site, on the other hand, does not give seniors a feeling of membership: To reach it, there are no codes or spaces reserved for members of the group. The site is open to all and by this fact it does not confer a feeling of membership. The listserv is exclusive to BEV Seniors. It makes members feel truly part of a tribe: Exclusive access, as well as the exchange of messages among BEV Seniors, consolidates their feeling of membership. During an interview, one of the members told me that she made a request on the listserv to find medical hardware to send to Chile. The other members helped her. That voluntary effort helped to strengthen links and give an identity to the group with respect to other communities and other populations. The fact that for some the listserv is the only active system to communicate with the other members of the group gives the listserv an important place in the construction of the community feeling of membership.

The genius loci of the BEV Seniors group

The grid

To assess the value of the genius loci in a network community, consider the following.

1. The immaterial dimension and the imaginary atmosphere that becomes a connection between community members. There are also the traces that members could leave of a physical/virtual object and space.

2. The memory, because without a collective memory a community doesn't exist.

The BEV Seniors group

"We are the first electronic village of the country."

From this assertion by one of the questioned BEV Seniors members, we can already see how the place becomes a link for the members of the group. They are proud to belong to this small town, a world leader in the development and diffusion of new technologies of communications. Moreover, when the media, national or foreign, interviews a member of the group, the other members are not jealous. On the contrary, they are proud to be part of the BEV Seniors group. People questioned by the media crystallize the positive energy of the members of the group, while strengthening the links between them and with the social environment that surrounds them. Moreover, the town government of Blacksburg practices a policy aimed at attracting retirees. The climate is favorable; there is very little crime; the town is peaceful; and senior citizens can generally afford to buy houses. Retailers offer discounts to senior citizens, and there are several associations that deal with the well-being of retired persons and organize many social activities.

> "... I think the community should be proud that [it has] this many seniors who are computer-oriented and who are working together. I think it's a good aspect of the Blacksburg Electronic Village."

Blacksburg is also magic insofar as virtual socialization can precede the real. One can become a member of a community before physically being part of it! Two of the people interviewed said that the listserv made it possible to be part of the community before even being on location. In particular, a couple said that before deciding on the place of their retirement they were members of the listserv. Once they had arrived in Blacksburg, they felt well-integrated in the social environment and knew very well the town, its inhabitants, its problems, and its moods. A similar case was noted by another interviewee. He said that he missed the city during a six-year absence and that the listserv enabled him to re-enter immediately in liaison and harmony with the community. The possibility of feeling part of the community, before even being physically part of it, makes this place and this group a little more magical.

> "... the real thing is the feeling of community—community that is generated—and I think especially for those people who have not been

here a long time, who are just moving into the area. At the last meeting we ran into a couple that had just moved here, and they're using [the listserv] as a vehicle to get to know people and to know the community. There's another woman who [has lived and] worked in the Washington area, but she's bought a lot and is going to be building and moving here in the next couple of years … she joined very early and [is reading] all the stuff she wants to learn about Blacksburg before she ever gets here. In fact, she did visit one of our meetings when she was in town about her lot and her land. And so, for different people I'm sure it gets different things for them. And really I think it's the opportunity to socialize with people of like interest … I think that's the biggest benefit."

This communication system, in particular, highlights the fact that in a community, being informed about what happens, and being able to send and receive messages on any subject, helps its members feel part of the community. Several interviewees agreed that being able to send messages and to answer and communicate with the group gives them a feeling of membership. Furthermore, they said that they feel more committed to all that touches the group and the community.

"The media are extensions of ourselves," said Marshall MacLuhan. The case of Blacksburg shows that media are true tools that extend body and spirit, while putting in harmony individual, tribe, and the larger community. In using the computer to manage a budget, plan travel, organize meetings, respond to questionnaires and shop, BEV Seniors are able to be more active and to be engaged at the center of the economic, social, and political life of their town.

This link is expressed with the construction of a double collective memory; one that relates to the group of retirees and another that relates to the integration of the group with the town. The listserv, in particular, recalls at every moment that there are other seniors who wish to remain active and to benefit from their retirement.

"… life gets pretty busy when you retire, let me tell you that."

The listserv allows not only the creation and the sharing of a collective memory for the group but also functions as a driving force that distributes energy to its members. The Nostalgia project aims to provide a collective memory of the town of Blacksburg. It is about rebuilding the social environment on the Web site.

Notwithstanding this effort, it is not simple to create a multimedia virtual memory shared by the members of the group. I asked whether after a trip, for example, the travelers could put the photographs of the trip online and write a story, but the answer was negative. They noted that their mastery of the technical aspects is not sufficient and that only the leaders of the group have the ability to use such tools as a scanner and HTML script and thus, to create a shared virtual memory.

> "I think probably a key point is that we all like Blacksburg; we all like where we live … [This is apparent] if you look on the Internet, on our home page, this nostalgia thing that I started. A lot of people wanted to share their memories and contribute to [the Nostalgia project]. And so, I think a large reason why [the Nostalgia project] was formed is that we all so like where we are."

The leader of the BEV Seniors group

The grid

To understand the value of the leader in a network community, consider the following.

1. The leader gives an imprint to the community, shares or retains information, establishes barriers, and facilitates meetings.

2. The leader's motivation and his or her personality can also define the community.

The BEV Seniors group

The role of the leader is fundamental. My interviews confirmed that there are several people considered to be leaders and that they are the heart of the group. Their leadership serves a double function. On one hand, leaders help members with problems of connection, use of software, etc., and they can solve practical problems that are fundamental to connecting and becoming part of the BEV Seniors group. On the other hand, the leaders function as charismatic figures in the group. Members speak about them not like people having political capacity in the group, but as people who are interested in the well-being of their community and who are always present to help. Interviewees appreciate the fact that in the event of a problem, someone can come to their house and help them leisurely, unlike a professional whom one pays.

> "I think that the main thing is being able to get help in an unhurried fashion."

By connecting people on the Internet, leaders also connect people to each other. The interviewees noted that for a completely abstract group like theirs the fact that there are leaders is fundamental, because without them the group would dissolve quickly.

Conclusions

As a conclusion I'd like to underline three fundamental points, described as follows.

1. The first relates to the organization of BEV Seniors group activities. I have already emphasized that the structure and organization of the group is very weak. Only recently was a steering committee established. The leaders of the group have always played a fundamental part, but without ever assuming the role of the "owners." Their wish is to share interests and passions with a community rather than forming and directing a group. The activities they organize make them feel in harmony with their community and with their own group of seniors. On this subject, I can point out the assumption of Maffesoli [1] when he speaks about energy characterizing social aggregations. There is that which can be defined as "outward energy" (ex-tendere) that tends toward the outside. A group uses this kind of energy to make finalized projects that exceed even the group itself. In addition, there is an "inward energy" (in-tendere), which tends toward the inside. This kind of group energy is not finalized with the achievement of a structured project, directed outside the group, but this

energy remains within the group and nourishes it. This seems to be the case in the BEV Seniors group.

2. The second point is related to technology itself. The members of the group seem to hope that technology is not an end in itself. Instead, they believe that it must be used as a resource and a means of being integrated with the social and economic life of the community. The facts show that the activity of the group goes well beyond the declaration of intent written in the BEV Seniors home page. The BEV Seniors do not limit themselves to learning how to use new technologies, but they are truly integrating the new technologies in their modus vivendi with the aim of gaining greater advantages in other areas. In addition, the computer is a formidable integrating factor in itself.

3. The third aspect relates to the use that BEV Seniors make of new communication technologies. I would define these users as "active observers" (voyeurs actifs). Everyone reads and appreciates the messages on the listserv. Everyone feels part of the group, thanks to the messages received, but few of them send messages.

The last point relates to the dynamics of the communication system that the BEV Seniors set up. Let us say that usually they use their listserv to tackle any kind of subject, including amusing subjects and everyday life (e.g., growing the best tomatoes and finding the best baker or mechanic). That, however, does not prevent them from using the system for other purposes. The listserv is a little like a weapon that they keep under their pillow to use, as needed, in case of problems. It is a very powerful barometer, a warning system ready at any moment.

In 1996, for example, Virginia Tech wanted to begin a project with General Motors, Ford, and Chrysler, on an intelligent road. The goal of the smart road project was to build roads with electronic sensors that allow communication and information exchange between roads and cars. The smart road construction constituted a political stake of some importance and directly touched many of the residents of Blacksburg. The members of the BEV Seniors group started to discuss the smart road on the listserv, each giving his or her point of view on the place of construction, on the possible problems, on circulation, etc.

> "Well, the smart road is a road that is going to be built out here ... Back when the county was having to take the land to build the road ... it got to be a big local political issue, and we had a lot of discussion about it on our listserv. Some people were for it; some were against it; some didn't give a damn. But nearly everybody had something to say about it ... one member of our, group ... is on the county board of supervisors who votes on all of those things, and so occasionally he asked for our opinion ... [for] more guidance on how to vote. This doesn't go on all the time, but whenever an issue of interest comes up we have a lot of to say about it."

This type of discussion and dynamic does not happen often, but this concrete case shows how if there is a political stake that directly touches the members of the group and the inhabitants of Blacksburg, the BEV Seniors have a communication medium, with a formidable power. The group's listserv, newsgroup, and face-to-face meetings constitute a screen for everyday life and affairs. The BEV Seniors do not use their group to engage in political action or to influence the decisions of the mayor. If necessary, however, they know that they have the possibility to do so and that they have enormous potential. The membership of the group, in synergy with the new media, represents a true soft system of homeopathic integration in the town and the community.

I asked interviewees whether they had widened their horizons and their circle of friendships. The general answer was positive, because the seniors met people whom they did not know before, or whom they knew only superficially (by sight). Since their membership in the group and their participation in face-to-face meetings, they can put a "name to a face"; they know their mutual interests, and they stop to exchange words when they pass on the street.

In response to being asked whether they engaged more in issues that interest them, the general answer was positive. Thanks to the Internet, they can look further into subjects that interest them. In addition, they widen their horizon, insofar as they take part in conferences or debates on subjects that are not directly linked to their own interests.

In response to being asked whether they are at least committed to the business of the community, the general answer was, even here, positive. Reading the online messages and knowing what occurs around them has made the BEV Seniors feel more committed to and concerned with their community. Being informed and reading the messages in the listserv has made them feel more responsible and concerned about problems that they would not have approached previously.

The research undertaken now enables us to say that BEV Seniors, and the dynamics analyzed between medium and users, makes it possible for the members to be fully integrated and active in their social environment or to reinforce the links between members of the group and other communities. These new forms of communication (that I consider to be new communicational environments) make possible the homeopathic integration between individual, tribe and larger community. The individual is better integrated with 1) the self because people who use new networked communication feel part of the world and find a place in society, with 2) the tribe because they can meet people with the same interests and passions, and they can establish with them new relationships and share emotions, and with 3) the community because people are more informed and involved in local and global issues.

Appendix: Basic Questionnaire

1. Membership (domus):

   - How did you first hear about the BEV Seniors?

   - What are the goals of the BEV Seniors?

   - What are the benefits of belonging to the BEV Seniors?

   - What are the obligations of belonging to the BEV Seniors?

   - What are the moments that "make people feel closer"?

   - What are some of the feelings you associate with being a member of the BEV Seniors?

2. Communication:

- How do you use the network to communicate with other members?
- Would you say that there are any shared codes among BEV Seniors?
- Do you send/receive information to/from the other members?
- Which media do you use (e.g., telephone, oral communication, Internet)?
- Do you send information online?
- Do you participate in building senior citizens' Web pages?
- Do you participate in the listserv discussions?
- How frequently do you send messages to the listserv on average (less than once a week, once a week, more than once a week, daily)?

3. Virtureal space:

- What are the BEV Seniors activities or events that you like most?
- What are the events online or face-to-face that you like most?
- Do you meet with other BEV Seniors—if so, when, where, and for what purpose ?
- How do you deal with difficult moments, crises, or problems with other members (online or face-to-face) of the BEV Seniors?

4. Place (genius loci):

- Have you built new social relationships with other seniors citizens members in Blacksburg?
- Have you met people from other communities?
- Have you built new social relationships with other seniors you have met through the Internet?

- Since getting on the Internet, do you feel more, equally, or less involved with people like yourself? (Why?)

- Since getting on the Internet, do you feel more, equally, or less involved with issues that interest you? (Why?)

- Since getting on the Internet, do you feel more, equally, or less involved in the local community? (Why?)

5. Leadership:

   - Who is the leader of the community?

   - What is his/her importance for the community?

   - What does he/she do for community members?

   - How would you describe her/his job, charm, character, and charisma?

Reference

[1] Maffesoli, Michael, *Time of Tribes,* Sage, 1996.

7

Networking Families into the Schools

by Roger W. Ehrich, Melissa Lisanti, and Faith A. McCreary

SINCE THE INCEPTION of the BEV in 1993, a significant effort has been made to incorporate the public schools in Montgomery County, Virginia, into the vision and framework of the BEV. Subsequent state funding encouraged confidence that the county's early progress in networked computing would lead to sustained infrastructure development. Once the initial excitement over the educational potential of networked computers in the schools subsided, however, state and federal sources of financial support quickly vanished, and it became clear that many schools and school districts such as ours would have to be financially and technically independent. Without state and federal initiatives, the national fate of network-based educational support in the schools began to slip ever more into the control of local politics. That, in turn, is dramatically increasing the potential for educational disparities from community to community, if not from classroom to classroom.

In March 1997, a distinguished presidential panel issued a report on the status and future of networked computing in the public schools [1] and laid out many of the hopes for the future as well as the financial implications of the vision it set forth. The panel vision was not so much a program to educate our children about computing as a program to integrate networked computing into the curriculum as a superior tool for facilitating learning. Undoubtedly, many of its expectations were based upon the success that many teachers were experiencing in using new technologies to support constructivist learning in the classrooms.

Others have also noted that one cannot dismiss the importance of broad-based network literacy in decisions to finance and support technology in the public schools. It is well argued by Slowinski [2] that the importance of Internet initiatives in the schools stem also from the need to develop human capital, from the role of information access and dissemination in a democracy and from the implications for personal empowerment. Recently Virginia Tech in collaboration with the Information Technology Association of America [3] estimated that there were 346,000 vacant information technology positions in the United States. The difficulty in finding qualified people to fill these positions was in part responsible for recent changes in immigration quotas to permit greater numbers of foreign nationals to fill these positions. At the same time, the United States suffers from gender equity problems, with women continuing to avoid information technology careers at a time when the potential rewards are as great as ever.

Just what are the financial implications of networked classrooms? The presidential panel report finds that currently just 1.3% of the quarter trillion dollar national education budget is targeted at classroom technology, with expenditures in 1994–95 of $3.3 billion. To provide a computer for each five students would cost $47 billion plus $14 billion per year for continuing support. The panel argued that professional development for teachers would consume 30% of the technology budget and estimated that giving all teachers just two hours per week to prepare lessons would alone cost $9 billion.

Such numbers cause understandable anxiety among school administrators who each year face increasing difficulties in raising enough support to solve the most pressing problems of leaking roofs, aging bus fleets, and declining real salaries for school staff members. In that context it becomes increasingly important to understand just what it is that might be achievable educationally through networked computing. Without statistics and hard evidence, it is difficult for school administrators to argue with their school boards and county supervisors that it is worth the political risk to quadruple technology expenditures nationwide as the presidential panel report recommends. While increasing a school budget is a sensitive topic in any community, publicity about the seamy aspects of computer networking and the recent report from the Homenet project at Carnegie-Mellon University that excessive networking activity increases depression and isolation [4, 5] make technology expenditures even more problematic.

Through a series of programs funded by programs such as the National Science Foundation's Networking Infrastructure for Education (NIE) and the Department of Education's Challenge Grant Program, a number of targeted research programs have been established to explore aspects of K–12 technology deployment. These include learning strategies, collaborative learning, digital libraries, staff professional development, evaluation methodologies, technology deployment, rural and minority education, network access, and technology integration. While these are all significant components of the larger K–12 technology integration process, it remains to be seen whether real long-term educational changes can be achieved, especially in view of well-established and easily costed techniques for affecting educational outcome, such as decreasing class size.

In 1996, Virginia Tech's Computer Science Department and the Montgomery County public school system initiated a joint project called the PCs for Families (PCF) program to determine whether, under the best of circumstances, immersive access to networked computing by both students and their families would affect long-term educational outcome. In this program, which is funded by the Office of Educational Research and Improvement (OERI) of the U.S. Department of Education, a fifth-grade classroom was designed at Auburn Elementary School (formerly Riner Elementary), with a networked computer for every two students. Virginia Tech is lending a networked computer for an extended period of time to the family of each student in the program. That enables students to work at home with their families in the same way that they do at school and to collaborate with one another and their teachers at any time. Networking and technical support is also free to participating families, so that physical access is eliminated as a barrier to education and family development. Our families need only make a phone call or send e-mail to receive on-site technical support. Furthermore, physical and personnel support is provided to the classroom teacher so that time and classroom environmental factors are eliminated as barriers to the educational intervention. Each year for three years a new fifth-grade class joins our program and upon graduation enters a normal sixth-grade classroom in nearby Auburn Middle School. We provide continuing networking and technical support as we track the progress and activities of our students and their families once they leave elementary school.

The classroom itself was designed using a participatory process involving project staff trained in technology, education, and design methodology working with the regular classroom teacher and the school principal. Since a new school was built and occupied during the third year of the project, this had to be done twice. The first classroom was in a double-width trailer with floors that could be penetrated easily to hide wiring. This provided a superior design compared with the new classroom which precluded floor-based wiring and had largely immovable fixtures around the classroom perimeter. In both designs the computers are placed on table clusters across the classroom where they are easily accessible to the children at any time. Such access supports constructivist learning in which activities are less structured and more spontaneous than in conventional classrooms. Two children share a computer and a mouse, but each child has a personal keyboard electronically multiplexed with its partner. The teacher has a custom-designed demonstration station with a ceiling-mounted LCD projector and a dedicated computer with special capabilities such as a scanner and shared ink-jet and laser printers.

Family involvement in education

It is well known that parents and families have a strong influence on the development of their children. Children spend much more time at home than in school. Even though specific skills and knowledge may not be acquired at home, civic, social, and learning attitudes are strongly influenced by the home environment, and these, in turn, affect the effectiveness of the school. Parental attitudes and values concerning schools, teachers, and education are rather transparent to children, and these influence classroom discipline and the pace and excitement of the classroom learning experience. Moreover, children are sensitive to the actual behavior patterns of their parents, and require consistent, sustained parental confirmation and reinforcement.

Considerable research confirms and reaffirms the importance of family involvement. Many of the studies up through the mid 1980s are summarized by Henderson [6] in a manuscript that makes exciting reading because it is so unequivocal. The studies cited are diverse, but some of them report even relatively short (one-year) interventions that show positive results on standardized tests. At this time there is little published literature on the effect of technology in the home, let alone long-term studies of its effects. Indeed the research dimensions are a bit overwhelming since there are so many new issues to be studied.

Gordon [7] identifies three principal aspects of parental involvement: home parental influence on the learning environment, school volunteer and governance involvement, and community involvement. The first and third aspects have been more thoroughly researched, but all show consistently positive results. Following Gordon, Henderson includes the following in her list of conclusions:

1. The family provides the primary educational environment.

2. Involving parents in their children's formal education improves student achievement.

3. Parent involvement is most effective when it is comprehensive, long-lasting, and well-planned.

4. The benefits are not confined to early childhood or the elementary level; there are strong effects from involving parents continuously throughout high school.

5. Children from low-income and minority families have the most to gain when schools involve parents. Parents do not have to be well-educated to help.

6. Involving parents in their own children's education at home is not enough. To ensure the quality of schools as institutions serving

the community, parents must be involved at all levels in the school.

7. We cannot look at the school and the home in isolation from one another; we must see how they interconnect with each other and with the world at large.

For this reason, we decided from the beginning to work with the families as well as with the children, reasoning that if there are no demonstrable results under these circumstances, one would have to question the efficacy of broadly based networked computing as a teaching and learning paradigm.

Since our children are in the classroom at most 30 hours per week, we decided that we would not use class time to teach them about computers. Instead each child spends an hour a week in a morning program to learn about technologies relevant to academic activities. Similarly, the parents are expected to attend an evening program one hour a week to acquire similar knowledge and to learn how to work with and supervise their children. This instruction is not provided by the classroom teacher but rather by project staff members. These include Melissa Lisanti, who assists the classroom teacher in technology integration and acts as a liaison between the school and the researchers; Markus Groener, who has primary responsibility for technical support; and Faith McCreary, who has primary responsibility for project evaluation.

Designing a technology-rich intervention

The design of the entire PCF program was driven by a macroergonomic perspective in which the design is viewed to have complex and interrelated environmental, technical, social, and organizational components. This structure makes it possible to view classrooms as systems having inputs, outputs, and a mechanism for transforming one into the other. The outputs involve the quality of student achievement, and we are interested in understanding how technology affects achievement and the various components of the classroom system. Student learning outcomes are dependent upon all of these components, and optimizing any of them may not be successful without optimizing all of them. Therefore, the design of the PCF program was complex and multifaceted. Super-imposed on the PCF design was the problem of designing a compre-hensive evaluation plan for the four components with both qualitative and quantitative aspects.

The organizational component

The organizational component concerns the functioning of the class-room, event scheduling, team organization, classroom policies and prac-tices, curriculum, academic standards, and authority structures. To the maximum extent possible, we tried to preserve the classroom teacher's responsibility and authority over these aspects of classroom protocol. One major change was made: In order to maximize the student's physical presence in the technology-rich classroom and exposure to the teacher's teaching style, it was necessary to reduce the amount of team teaching and student rotation through the classroom. This artifact was the result of having resources to support only one technology-rich classroom and one teacher.

The technical component

The technical component includes the educational technology and the pedagogical techniques it supports; almost every aspect of this compo-nent was redesigned, including the furniture, the room layout, the technology, and the teaching style. We began by having one of the project principal investigators, a master teacher, work closely with the classroom teacher to provide mentoring and encouragement for the adoption of practices to support constructivist-style learning. That entailed giving the students more autonomy in their learning activities, encouraging indi-vidual exploration and initiative, lecturing less to the class as a whole, and integrating the new productivity and communication tools provided by networked computers. To support the teacher in making these changes we provided a technical support partner who could assist the teacher with the technology, with technology integration, with classroom activities, with the student technology training sessions, and with student and family communications that were stimulated by network access.

The basic classroom concept was that the technology was to be forever present, though not obtrusive, available just like any other classroom tool such as a book or a pencil. That required redesign of the classroom layout to make the computers easily accessible without obstructing viewing or other work activities. We found that as a result we had inadequate space for group activities such as story telling or discussion. Since there was one computer for every two students, collaboration between partners was encouraged naturally. While this worked well, we found that partners had to be switched frequently during the year as social and work relationships changed.

Last but not least, we developed a physical support structure for the technology so that the network, computers, displays, and printers would work properly when they were needed and so that supplies, software, and consumables would arrive on time, both for the classroom and for the home computers.

The social component

The classroom social subsystem consists of students and teachers and is characterized by beliefs, attitudes, motivations, and patterns of social interactions. While the teacher may affect certain patterns of interaction by controlling student placement and room configuration, observing the various aspects of the social component provides us with major clues about how learning is taking place in the classroom.

The environmental component

The environmental component consists of the larger school organization and the family. Every teacher and classroom is housed within a complicated organizational structure that encompasses the school, the school division, its governing bodies, and the associated community. These bodies normally form the teacher's support structure and greatly influence a teacher's ability to organize and maintain the physical classroom and computing infrastructure, to maintain discipline and interpersonal relationships with school staff members, and to take risks with new ventures such as the PCF program. By working closely with the school principals and by supplying the necessary financial resources, we were able to ensure strong support for the classroom teacher.

The teacher's relationship with the school principal is critically important to the classroom process because the principal is ultimately responsible for establishing behavior norms and bounds and for resolving conflicts. In the same way, the principal also has strong influence on the relationships between the teachers and the families and in many cases may affect family attitudes toward the school system. The needs of our project often conflicted with district operating procedures and constraints. For example, in our classroom, educational perspectives and communication activities were very much outward-directed to give the students maximum awareness of and contact with the larger world around them. The school system, on the other hand, as a public agency, perceives the need for accessibility constraints and accountability through filters, firewalls, and intranets. In such situations the role of the school principal as a mediator has always been extremely important.

Last but not least, the family is a strong driving force that is a prime determiner of student motivation and attitudes toward the school, the teacher, and the learning process. For this reason we have made a concerted effort to engage all family members in the PCF program, to encourage the parents to work with their children, to explain the ethical structure we are laying out for their children, and to convince them of the value and joy of learning.

Evaluation

The original project goals were very specific: to see whether in a three-to-five-year period we would find changes in standardized test scores between our children and those in a matched control group selected anonymously from children across the school district. Once the project started, though, we realized that we could learn much more about the dynamics of the home-school system and began the design of a much more extensive set of evaluation procedures. These began with yearly interviews of all the project parents and project children. A growing regimen of survey instruments was designed and administered to detect changes in behavior, attitudes, learning practices, and family relationships. We collected evaluations from all project personnel who knew the children or the families, and we administered both learning-style and personality-typing (Murphy-Meisgeier) instruments to the children. Both children and parents conduct their e-mail and Web communications through our proxy server, which is logging their activities. All project e-mail sent or received by project staff is archived so that all communications are preserved and a fairly good project activity log can be reconstructed. Similarly, all use of project-designed collaboration software, such as shared editors and chats, are logged and analyzed.

In our project we are also extremely interested in knowing what happens in the classroom and what changes are taking place in the ways in which the classroom teachers use the technology. Activities on selected computers in the classroom are being logged to give us detailed information about their use. Videotapes are being made so that we can log classroom movements, lessons, student collaboration, and teaching style.

Teaching in the PCF classroom

A goal of the PCF classroom is to provide a technology-rich and dynamic learning environment in which students have the opportunity to track stocks as passionately as stock brokers, to feel the same sense of adventure as pioneers on the Oregon Trail, and to become biologists, oceanographers, or civil war soldiers. These kinds of experiences empower students to educate themselves with the assistance and guidance of the classroom teacher. So that this kind of constructivist learning environment can be most effective, it is imperative that students have opportunities to pose questions, to explore, and to guide themselves so that the curriculum becomes meaningful and relevant to each student. Technology provides powerful opportunities for students to incorporate these learning strategies, making it a natural complement for this kind of environment. It provides motivation for the children to recover some of the natural inquisitiveness that many have already lost at this early age.

Students come into the PCF classroom at the start of the school year with fairly traditional experiences in education. Many have had some exposure to constructivist environments and possess limited computer skills acquired in previous grades. However, the learning environment in the PCF classroom is, in certain fundamental ways, unlike any classroom that they have been in before. While the curriculum remains the same for all fifth-grade classrooms in the county, the processes by which students master the curriculum are often new in our classroom because the accessible technology provides a wide variety of learning strategies previously unavailable to students. The computers are no longer used for drill and practice, as one of a group of learning centers, or as a rotational station for a few students at a time. CD-ROM use is limited mostly to the occasional encyclopedia reference in the PCF classroom and a small library that students can check out and take home. Instead, spreadsheets and databases add new dimensions to organizing and analyzing information; network technologies stretch the traditional boundaries of group work with a wide variety of collaborative facilities; and presentation software allows students to design projects that require not only research and writing, but also integrating their ideas with sound and images to create cohesive and effective oral presentations. With these tools available to them, students are eager to participate actively in their lessons, and their enthusiasm is contagious.

With the assistance of the classroom teacher and technology specialist, students integrate technology into the daily activities of all subjects. Early in one year, students navigated through the Internet to discover why people were willing to undertake the 2,000-mile journey to Oregon in the 19th century. Students began their searches with a Web site designed by the technology specialist that mentioned a few of the hazards on the Oregon Trail, posed a few preliminary questions to generate interest, and set them up for the task of finding some answers. A few well-chosen Internet links got them started on their journey. They traveled at their own pace to Web sites that included a primary source diary, lyrics to songs, and a historical center on the Oregon Trail, so that they could piece together a picture of life on the trail and the motivations for going west. They chose what direction to take and what sites to concentrate on, and they were free to follow other links from those sites to find more information. Because students worked through these materials themselves, the ideas stayed with them. Their knowledge and understanding of trail life was later evident in diaries that they wrote from a settler's perspective on the Santa Fe or Oregon Trail. Many of the student papers reflected an understanding of the difficulties that settlers faced, the reasons why people chose the trail, the style of travel, and an appreciation of the hardships that the journey entailed. This exercise marked the beginning of the students' own journeys of exploration into the Internet that continue throughout the year and, it can be hoped, into the future.

It is a journey as fraught with peril as that on the Oregon Trail was for the early settlers, but because the technology is so accessible for all students in both the classroom and home, students have extraordinary opportunities to learn how to use the Internet as a resource both wisely and safely in the context of curriculum-related topics. Over the last two years, questions and discussions have evolved into a kind of critical literacy curriculum that must occur simultaneously with Internet research. Some activities require students and parents to evaluate Web site content and presentation, students record Web site addresses in project logs when they use the Internet for research, and teachers give lessons on critical literacy that advise students and parents to corroborate information and to use Internet research cautiously. Because these valuable lessons occur behind the scenes of projects ranging from reptiles to geography, instruction achieves a two-tiered result: curriculum mastery and an understanding of the deeper issues surrounding research. By involving parents in these lessons, we hope that these research skills will continue to be reinforced once students have left the PCF classroom and must use the Internet independently.

The Internet is just a small piece of the technology utilized in the PCF classroom. We use a variety of technologies to spark interest and attract student participation. One class used a spreadsheet to track the activity of stocks that students chose to follow. During class students used the Internet to choose their stocks, employed math skills to calculate how much stock they could afford to buy with a predetermined amount of money, and entered simple formulas to calculate change in performance. The enthusiasm generated by the activity sparked student discussions on the millionaires of the early 20th century, life in the 1920s, the stock market crash, and the beginning of the Great Depression in the United States (another piece of the fifth-grade curriculum). The classroom teacher and the technology specialist were inundated with questions. In this hectic and inquisitive environment, we not only engaged all students in the activity, but we observed the students cross the curriculum from math to social studies on their own initiative.

This jump was not planned as part of the lesson and was as much a surprise to the classroom teacher and technology specialist as it was to the students. However, in the PCF classroom, the teacher must be flexible in accommodating these transitions. In fact, after observing student behavior during several lessons when these kinds of jumps occurred, the students are excited about directing their own learning and, most importantly, taking their teachers with them. The teacher's surprise adds to their sense of adventure and exploration. In these circumstances, the role of the classroom teacher must move toward facilitator and away from lecturer in order to maintain that spark of student interest that is so critical in achieving student involvement and motivation. This transition of the teacher's role is a cornerstone of constructivist learning environments.

Because the atmosphere in the PCF classroom encourages student initiative and creativity, many students gained confidence in their abilities to learn and to contribute to collaborative activities. At the end of the second year of the project, a group of students took control of a comprehensive project to describe their experiences in the fifth-grade. Eight students became the editors of a presentation and magazine that the entire class contributed to, including one student who had started the year very isolated from peers. Article ideas, the project title and mascot, software choices, and organization were all contributed by students. Editors were genuinely excited to participate in working lunches during the school week to design the project and most often wanted to give up recess time as well. The project's private chat room was the setting for brainstorming sessions by editors after school hours, sometimes involving the technology specialist and other times not. Network technologies leveled the playing field for all students that wanted to participate, including the isolated student mentioned above. Perhaps most importantly, by the end of the project, students were considering ways that they could have improved their final product, deciding what they would change if given another opportunity, and reflecting on what they'd learned about project organization from this experience.

In this kind of dynamic and sophisticated classroom environment, it would certainly be understandable for parents to feel inadequately prepared to deal with the new technology in their homes and to help with their children's homework. Students often take home assignments for the Internet, word processing, spreadsheets, databases, and a wide variety of technologies. The PCF project includes parents in this transition by offering an evening program for them so that they may be more aware of what their children are experiencing in school. Network technologies at home also give parents and teachers the opportunity to communicate with each other at their convenience. E-mail updates from the classroom let parents know what kinds of activities are going on at school. We use communication wherever possible to keep parents involved in their children's educational experience in the fifth grade. Some parents have become quite adept with communicating via e-mail; others continue to write a note to the teacher in all circumstances.

We also use the evening program to bring students and parents together. Doing activities on, for example, listening skills, following directions, or designing a holiday letter, we encourage students and parents to share time and technology with each other. For some of the parents, that evening session was the first time that they had used the computer to work with their children. Many families began the joint sessions showing signs of awkwardness and reservation. They were very quiet; conversation between parent and student was minimal; and they were reluctant to turn on the machines in the classroom. In contrast, by the end of the session, students were telling parents what to do; parents were sharing their enthusiasm; and everyone was interested in solving the puzzle. One student who had come with his father asked for copies of the puzzle instructions to take home to work on with his mother, who had to that point been almost completely uninvolved with the project. Maybe these kinds of opportunities will encourage families to work together and involve parents more intimately with their children's education.

The PCF classroom is a constructivist learning environment for the teacher and technology specialist as well as for the students and parents. We've dealt with challenges like stimulating creativity and teaching students how to ask questions. Many students at the beginning of the school year were not ready for the less restrictive constructivist environment and were unprepared to take responsibility for their learning. Without a strict structure to which to adhere, they were lost and inefficient. Those circumstances necessitated modifications in our learning environment, and we had to scaffold activities to gradually bring students into the constructivist mind-set. Each year brings a new group of students, and we change directions, modify approaches, and set new goals. Future challenges will include emphasizing the polished finished product, placing an even stronger emphasis on critical literacy, and experimenting with new collaborative technologies. Traditional instructional techniques like debates, role playing, or even direct instruction are not completely abandoned in the PCF classroom—just balanced and integrated with constructivism and technology to add motivation, spark interest, and achieve even more meaningful learning for every student. While the impact of immersive technology on long-term educational outcome cannot yet be determined, comments of the fifth-grade students indicate that they enjoyed their new learning environment immensely and were reluctant to leave it for a more traditional one in the sixth grade.

Project results

It is difficult to present a simple analysis of the way in which the PCF program has affected the participating children and families. There seem to be few stereotypes and even among individual families, substantial complexities.

As one might expect, the more talented among our children were delighted by the new challenges and tended from the outset to enjoy substantial family involvement and support. As other researchers have pointed out, these children may have had less to gain because they would probably have been successful, regardless of their specific experiences in fifth grade. However, there were other children such as Judy, an average child from a difficult family, who in our program found role models and reached out through her new communication channels to the world beyond. Similarly, there were children like Max, academically detached from his classmates, who found that he was especially good with the technology and began to thrive through the attention and peer recognition he received.

We have also had in our classrooms a number of children, primarily boys, who were disengaged learners who could not be reached even with the very substantial motivational power of the technology. They tended to be visual learners who seemed to do well only when intensely stimulated. At the end of the year, it appeared that these children were beginning to make progress, and the teachers regretted that the actual intervention ended there. Their conclusion was that somehow these children had lost their curiosity and self-motivation and that being suddenly placed in a constructivist learning environment was wrong for them; they needed much more structure than our classroom could provide.

As long as our proxy server has been recording data we have noted that the children and the families are very active on the network when they first install their equipment at home. Subsequently, use declines steadily as the novelty wears off. We also note that the children perceive a decline in parental involvement as the year progresses, while the parents report the opposite: that their support activities are constantly increasing. A plausible explanation is that the children learn so much faster than their parents that their expectations of their parents grow much faster than their parents' abilities.

As the first set of parent interviews were completed, we also learned that in a number of our families, both parents held jobs involving long hours or hours that made it difficult for them to spend time with their children. For example, in one family it was frequently the case that neither parent would come home before 9 P.M. on weekday evenings. Thus, sometimes job stress interfered significantly with parental involvement, especially in lower income families that, in principle, had the most to gain from a program such as ours. In other cases, one parent who had the time would become the active parent, while the other one, who was job-bound, would withdraw. Understandably, single parents often found themselves under even more job stress.

The PCF program was designed with the expectation that the parents would provide continuity when the children moved back into a normal school setting, so that the skills, attitudes, and habits they had acquired would be transferred to their new sixth-grade environment. However, we are learning that in many cases the lack of school support, technology, and reinforcement in middle school canceled much of the progress that we had made. Thus, we believe that technology interventions such as ours should be made system-wide and not left to individual teachers and individual classrooms where they might have only temporary effect. There is also ambiguity in teachers' attitudes toward parental involvement. It is common—for example, see [8]—that many teachers espouse home-school relationships while at the same time avoiding perceived interference from parents with the classroom environment and control. Needless to say, we are strong advocates of moderating those barriers and using the technology to encourage much closer parent-school interaction and collaboration.

From the family interviews, we are learning that the families are indeed adopting network-based technologies to serve family needs. For example, they are searching for information, using the Internet to plan family vacations and activities. Some parents are now designing presentations for their jobs. Many families and children are now active communicators through e-mail. Some parents, though, report that working with computers and with e-mail is so difficult that they make little use of their capabilities and have forgotten much of what they learned during the previous year. It appears that some aspects of the degree of family technology integration is explained by what is known in the communications field as critical mass theory.

Critical mass theory [9] seeks to explain the diffusion and acceptance of interactive media such as telephones, paper and electronic mail systems, and electronic conference systems within communities. Key to acceptance is the idea of universal access; for example, the telephone has extreme value because it can be used to access almost anyone within the United States even though the average telephone owner would actually use only an increasingly small part of that capability. When a new technology such as e-mail becomes available, its acceptance by any person or party is determined first by need or perception of value gained by such access. One determiner of such a perception is the acceptance and willingness to respond on the part of other adopters of the same technology. Other determiners of acceptance are the costs of the technology, the skills required, ease of use, the literacy and training on the part of the adopters, the time required to use the technology, the perceived gains from using it (which also relate to the size of the accessible community), and actual usage costs. For any technology to succeed at all there needs to be a community of early adopters whose perceptions of value are so strong that enough of a community is formed to induce others to adopt later.

In the case of the PCF program, the buy-in costs and usage costs are zero, and training and support is universally provided. Moreover, both parents and children are specifically asked to read their e-mail and to be responsive; in other words, our families are coerced into adoption for a one-year period. During that time some of our families learn that extended family members in remote locations are now easily accessible or that they can easily carry out job-related functions. Similarly, many of the children begin communicating with their peers about an increasing variety of topics. However, in other families, there is just no time; keyboard or English literacy is insufficient; or the extended family is local so that the telephone is a more logical choice of medium. Thus, after the first year of the program, after most families have had the opportunity to assess the technology in the context of their own family culture, and after we remove the immediate project-related needs to communicate, we begin to see the real effects of the technology on the families.

One of the problems that has been repeatedly brought to our attention is that the middle school teachers are not accessible by e-mail. Thus, a number of the families who looked to the technology as a way to support closer home-school communications were actually rebuffed by the school system. We also find from the interviews that parents sincerely want to support their children and to give them opportunities for achievement that they themselves did not have. However, family constraints and resources vary, and the native parental abilities to provide support vary and play a great role in determining the way in which computers are being integrated into family culture.

The PCF program has adopted new strategies to respond to what we have learned. We are now working more closely with middle school teachers to increase their capabilities and their supportiveness. We are working more with the families to show them ways in which they can work with their children; as noted earlier, once a month we ask that the parents come to the evening class with their children, and that has been well-received. We are also beginning to receive requests from our first group of families for additional training, and from some who were apparent nonadopters, for technical support. Thus we are seeing signs that longer-term changes in family culture are indeed occurring. Our original hypothesis was that some of the important changes would be long-term changes, and that appears to be the case. However, it will take much more work and more time to learn of the effects of the new network-based technologies on the dynamics of the home-school system and to help the K–12 educators to capitalize on the gains.

One of this year's project interviewers now ends each parent interview with an open invitation to add comments. Almost every parent is effusive about the project and positive about the changes it has brought to the family. In many cases the parents point out that it would have been absolutely beyond their means to acquire a networked computer on their own. These are rewarding things for us to hear, but it still remains to show, feelings and perceptions aside, that there has been a quantifiable effect in some aspect of their life or in the achievement or aspirations of their children. These results will take additional time to reveal, but there are indicators that the program effects will be significant.

References

[1] Shaw, David E., et al., "Report to the president on the use of technology to strengthen K–12 education in the United States," President's Committee of Advisors on Science and Technology, Panel on Educational Technology, March 1997, http://www2.whitehouse.gov/WH/EOP/OSTP/NSTC/PCAST/k-12ed.html.

[2] Slowinski, Joseph, "Internet in America's schools: potential catalysts for policy makers," First Monday 4 (1), January 1999, http://131.193.153.231/issues/issue4_1/slowinski/index.html.

[3] Virginia Tech Division of Continuing Education, "Help wanted 1998: a call for collaborative action for the new millennium," Information Technology Association of America, March 1998, http://www.itaa.org/itaapub.htm.

[4] Kraut, R., W. Scherlis, T. Mukhopadhyay, J. Manning, and S. Kiesler, "The Home Net Field Trial of Residential Internet Services," *CACM* 39 (12), December 1996, pp. 55–63.

[5] Kraut, R., M. Patterson, V. Lundmark, S. Kiesler, T. Mukophadhyay, and W. Scherlis, "Internet Paradox: A Social Technology That Reduces Social Involvement and Psychological Well-being?" *American Psychologist,* 53 (9), 1998.

[6] Henderson, A.," The evidence continues to grow—parent involvement improves student achievement," Columbia, MD: National Committee for Citizens in Education, 1987.

[7] Gordon, I., "What Does Research Say About the Effects of Parent Involvement on Schooling?" paper for the Annual Meeting of the Association for Supervision and Curriculum Development, 1978.

[8] Lightfoot, S. L., *Worlds Apart—Relationships Between Families and Schools,* New York: Basic Books, Inc., 1978.

[9] Markus, M. Lynne, "Toward a 'Critical Mass' Theory of Interactive Media," *Communication Research* 14 (5), 1987, pp. 491–511.

8

Managing the Evolution
of a Virtual School

by Roger W. Ehrich and Andrea L. Kavanaugh[1]

Context for a virtual school

From the beginning of its operational phase in October 1993, the BEV has engaged the local school system in discussions and planning to promote network resources for education. Montgomery County, Virginia, has diverse demographics: at one extreme, a nucleus of computer-literate, network-trained individuals typically associated with Virginia Tech and businesses in Blacksburg; at the other, rural residents without telephones who have only a limited sense of connection to an information-based society.

The population of the county in 1995 was estimated at 74,000, with an unemployment rate of 4.5%. Overall population growth rate for the 10-year period ending in 2000 is projected to be 7.7%, about 67% of

1. The authors are grateful to Larry Arrington, Herman Bartlett, John Burton, John Carroll, Andrew Cohill, and Sally Laughon for their contributions and comments.

the expected statewide growth rate. The greatest population increases are estimated for the age groups 45–64 (37.5%) and 65 and older (20%). A decrease of 1.4% is projected for the population in the age group 20–44 [1].

The Montgomery County school system in 1995–96 consisted of 20 schools with 670 teachers and professionals, educating roughly 4,400 elementary, 2,100 middle school, and 2,500 high school students at a cost of $5,433 per student, or roughly $49 million. To cope with new facilities and technology, a per-student budget increase of 11% was requested for 1996–97. Total student enrollment is expected to increase 8.4% between 1995 and 2000. Ninety-eight percent of the school district's grade 12 enrollment graduates from high school. Approximately 71% of high school graduates are reported to continue education beyond high school. Of the county's K–12 students, 28.1% currently receive free or reduced-price lunches.

At the outset of the operational year in October 1993, one full-time technician was responsible for all electronic technology, including networking and computing. The schools did not participate in the BEV beta test phase (January–September 1993), but they began to access the Internet via the BEV at the outset of the operational phase in October 1993. Bell Atlantic generously agreed to share the cost of connectivity for schools in the rural area of the system. All schools were offered SLIP connectivity over existing phone lines into the modem pool managed by Virginia Tech's Communications and Network Services, owner and operator of the university cable plant and services (television, voice, data).

Early objectives

The Internet was not new to teachers and administrators in Montgomery County when the BEV first offered them connectivity. Like other school districts in Virginia, they had had Internet access for several years via a state-supported system, called *Virginia Public Education Network* (VAPEN, pronounced "Virginia PEN"). The VAPEN system was a regionally managed network with low-speed access. Thus, in 1993, a number of teachers had limited familiarity with the Internet. While classrooms generally did not have telephones, about 40% of the district's classrooms had computers with dialup access to VAPEN via a shared phone line or local area network (LAN). The schools' computer inventory was a mixture of Apple IIe vintage machines, Windows-less PCs, and low-end Macintoshes. With the exception of a few targeted programs, such as business management, writing to read, and keyboarding, few computers were used in regular instruction. Most teachers who had computers used them primarily for school and class administration and for running drill-oriented programs.

According to a district-wide technology survey, on average 18% of the teachers had access to the Internet (Figure 8.1), with a significant number of teachers indicating that they wanted access and training [2]. Roughly 31% of the teachers used e-mail for professional communication, and about 13% used it for instruction (Figures 8.2 and 8.3). One of the teachers attributed the early adoption of networking to one of her colleagues, formerly the administrator of the Radford node of VAPEN, who made a concerted effort to train as many teachers as possible about e-mail and basic networking. Local educators' knowledge and familiarity with networking via VAPEN made the task of training for the BEV much easier, because a basic cultural tradition had already been established among the e-mail users. Those with no background in networking had more to learn. The mechanics of using e-mail are relatively easy to learn. The conceptual integration of e-mail into personal and professional life takes a bit longer, at least until the new user establishes a group of peer correspondents and learns how that "community" can improve productivity.

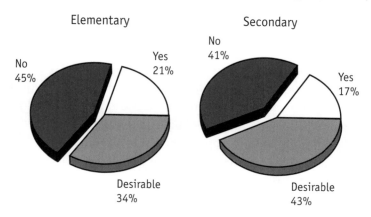

Figure 8.1 Do you have a connection on the Internet?

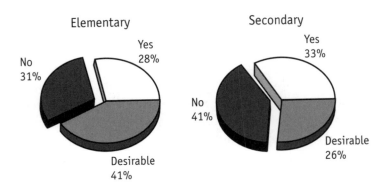

Figure 8.2   Do you use e-mail to communicate with collegues?

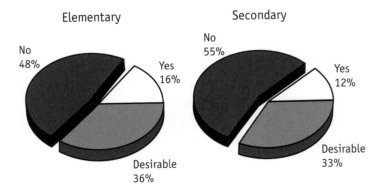

Figure 8.3   Do you or your students use e-mail for instruction?

Among the early (Fall of 1993) objectives was to provide uniform access across the district, in all of Montgomery County, not just in the town of Blacksburg. One BEV staff member was designated as full-time liaison to the schools to install the BEV front-end software on school computers and to help train teachers, administrators, and students, either one-on-one or in small groups. In addition, BEV staff held seminars and demonstration classes for school board members, the coordinators of the gifted-student program, and other education administrators and interested teachers to familiarize them with the benefits of high-speed networking and to enlist their support in promoting the use of network services in the schools. Those hands-on workshops, where administrators and planners got to play on the World Wide Web with high-speed connectivity, were more effective and persuasive than any amount of discussion and description offline. The workshops had to be conducted in laboratories on the Virginia Tech campus, because the county schools had no Internet-connected laboratory until the spring of 1996.

The schools view networking as a means to provide a much more meaningful educational experience to the students by giving them access to current global information and a way to conduct critical research and to collaborate with peers, parents, and mentors throughout the world. It is generally acknowledged that increased community involvement improves K–12 education [3–5]. Set within the BEV, the Montgomery County school system is accessible to all students and teachers, as well as parents, mentors, and other interested parties in the county. The schools also see networking as a way to provide students and families across the county with equal educational opportunities and to reduce disparities among schools, something that has been difficult to achieve given the county's rural-suburban nature.

The school district is exploring a variety of network services, including cost-effective videoconferencing, to bring educational opportunities to all students, particularly to those who are the most isolated and geographically remote. While the local school system has made substantial investments of capital and human resources, other realities of public education sometimes shed a harsh light on aspirations for change. In the first operational year (1993–94) of the BEV, most of the problems that teachers had were associated with the limitations of dialup access and with the installation of BEV software on old computers. Planners from the BEV staff, Virginia Tech faculty, and the school system were quickly convinced that high-capacity computers and high-speed connectivity would be required before most teachers would integrate Internet resources on a routine basis. The school system created a new position, supervisor of technology, and hired an experienced and knowledgeable individual to lead the schools in upgrading their network infrastructure, training, and services.

Engaging the schools

In the fall of 1994, the Montgomery County school system established a committee of teachers, administrators, and local citizens to draft a technology plan for the schools. The impetus for the plan originally came from the opportunity to receive state funds for library automation, which required having a technology plan on record with the Virginia Department of Education. That plan is available online in the Montgomery County Public School's Web pages at http://www.bev.net/education/schools/admin/technology.html.

The technology committee for the school system made an initial 1995–96 budget request of $1.6 million. Of that total, only $403,000 remained after negotiations with the school board, which was requesting a $3.3-million increase, or 16.7% over the previous year. If invested only in computers at $2,000 each, the entire technology budget would provide fewer than a third of the professional staff with a single computer each! Unfortunately, the scarcity of funds is typical for a rural U.S. school. According to the 1994 technology survey, only 5% of the teachers indicated that they would be satisfied with only one computer (Figure 8.4). Figure 8.5 shows that, on average, 73% of the teachers wanted computers both in labs and in their classrooms.

One important budget item that the school board approved for funding was the establishment of two additional district-wide technology coordinators. The technology committee recommended that those positions should be increased to one per school over a five-year period. Thus, in the 1995–96 school year, two new full-time technology

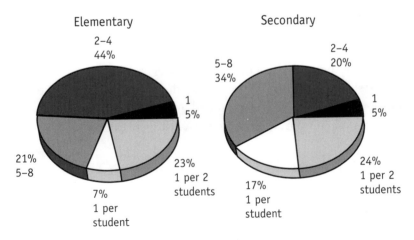

Figure 8.4 How many computers would you prefer in a classroom?

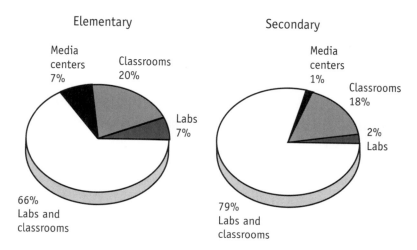

Figure 8.5   Where would you like to see student computers?

coordinators (who were also experienced teachers) began to assist teachers throughout the county with software and hardware, information, training, and troubleshooting. In addition, the school board approved a second full-time technician to assist with infrastructure building and maintenance in the 20 schools.

At the same time, the state-supported Internet access system, VAPEN, centralized its technical and operational management at the state capital in Richmond in an effort to reduce costs. To further cut back on expenses, VAPEN restricted access to its modem pool to one hour of use daily per user. At high-load times, no user was guaranteed access either by network or by modem pool. The concurrent alternative availability of high-speed networking (via T1 transmission lines at speeds of 1.54 Mbps) was very attractive to the school system. That alternative access option became possible for six schools with support from Bell Atlantic-Virginia through waived transmission charges and the National Science Foundation (NSF).

In September 1994, the NSF awarded Virginia Tech's college of education, department of computer science, and information systems (departmental home of the BEV) a planning grant in the amount of $100,000 to work with the county schools to plan the future of network-based education and a network-based virtual school. In concept, a virtual school is an unbounded educational environment with no walls, no halls, and no bells; where (virtual) collaborative classrooms encompass the entire community and exploit connections among diverse educational resources: schools, libraries, homes, businesses, government, local and global networks, and individuals. The participants are working toward an educational system that extends beyond the physical school into the community, in which both parents and children continue to communicate, to learn, and to collaborate in a rich network environment.

The NSF planning program, entitled "Planning for Virtual Schools in Electronic Villages," was undertaken in collaboration with educational content providers (e.g., Busch Entertainment Corporation and Scholastic Network) and networking support from Bell Atlantic-Virginia. In the planning year, a pilot program was initiated to create an emergent virtual school that comprised teachers and classes in two elementary, two middle, and two high schools, interconnected by T1 transmission lines donated by Bell Atlantic-Virginia. A number of research projects were planned or under way with individual teachers at several of those schools, but modest equipment purchases made possible by NSF came at a crucial stage of the projects. The equipment included seven Macintosh 6100 systems for connection to Ethernet LANs in the schools. In addition, seven used 80386 systems with modems were purchased for teachers to use at home in preparing lessons.

The planning team consisted of representatives from the schools, Virginia Tech faculty, and the BEV staff. Representatives from the schools included the technology supervisor and several coordinators of the gifted-student program. Networking is not targeted at gifted students; in fact, it appears to be equally successful among attention-deficit–afflicted and less talented students. However, the gifted-program coordinators are among the most experienced with the integration of network resources and communication into the curriculum and have more freedom to explore new educational technologies. They have been the most active in training other teachers, especially in the early years. Representatives from the Virginia Tech faculty included professors from the computer science department and the college of education (principal investigators [PIs] on the grant). Representatives from the BEV included the director of research (also a PI on the grant) and the director of the BEV.

Selection of participating sites for the pilot virtual school was made on the basis of the location of motivated teachers and geographical distribution among rural and urban (suburban) areas of the county. One of the rural school strands (elementary, middle, and high) is characterized by close geographical proximity; the Auburn middle and high schools are located in the same building, and the Roner elementary school is only a few hundred yards away. Selecting Auburn as the rural site made it possible to gain economies of scale. Bell Atlantic provided one T1 line to the middle/high school, and the schools paid the additional cost of running fiber for a few hundred yards to the elementary school. In that way, cost-effective high-capacity network access was provided to all three schools (at all grade levels), rather than to just one rural school at one grade level. In the town of Blacksburg, one elementary school, the middle school, and the high school were the three counterparts to the rural sites.

The schools selected 12 teachers, two from each of the six schools, on the basis that they were the most motivated and interested in network applications for education. Those teachers were eager to participate in a series of workshops intended to familiarize them with advanced network resources and services and to participate in future planning of a virtual school. The grant supported the release time of the teachers to participate in planning and training meetings. The intent was that participating teachers would become mentors for other teachers, a model that seems to have worked well.

The BEV provided a series of training sessions for those teachers to work with network resources and services and provided follow-up one-on-one assistance from Virginia Tech faculty, graduate students, BEV staff, and the school's own trainers (mostly coordinators of the gifted-student program). Progress was slow at first, but there was a sudden burst of intense activity when the teachers overcame hardware and software problems related to accounts, passwords, and physical connectivity. Soon those 12 teachers had mastered *hypertext markup language* (HTML) and began publishing school information on the Web, lessons for their students, and student work, mostly on evenings and weekends.

Because the schools had no laboratory computer facilities, teachers began to take their classes to the New Media Center at Virginia Tech (a public networked computer laboratory, provided largely by Apple Computer), where they could work on network-based lessons. Two teachers worked with their students on a networked multimedia magazine about life in Montgomery County. On teachers' workdays, minicourses were offered on computing, networking, and HTML. By the end of February 1995 (halfway through the planning year), the first school had a formal presence on the Web, along with the central administration. By the end of the planning year (September 1995), most schools in Montgomery County had Web pages (Figure 8.6). The effort was managed by the technology supervisor of the school system and by motivated teachers on evenings and weekends. In fact, the success of the program so far has surprised and pleased both grantees and teachers (see http://www.bev.net/education/schools/index.html).

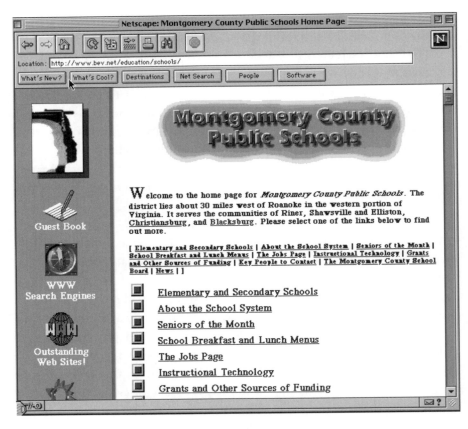

Figure 8.6 Montgomery County Public Schools (MCPS) home page.
©1997 Netscape Communications Corporation.

The planning period provided the basis for subsequent training and diffusion of network resources among the Montgomery County schools, with support from the school system, the state, and outside grants. Most notably, the NSF awarded a subsequent grant to Virginia Tech's computer science department and the Montgomery County schools for the development and testing of software to support virtual collaborative science laboratories. The TIIAP of the U.S. Department of Commerce awarded a demonstration grant in October 1995 to support, among other things, scaled-up training and the integration of network resources into the curriculum throughout rural schools in Montgomery County.

All training materials and other documentation associated with the local K–12 education effort are being posted online as part of a larger clearinghouse effort, so that teachers, trainers, and other educators can benefit from those materials regardless of their physical location. The effectiveness of transferring skills and knowledge in that format is being evaluated; it is expected that those materials will be most effective when used in combination with local training efforts and access to hands-on computer laboratories. Providing training documentation and guide-books to trainers and teachers should reduce the overall cost for other groups conducting local training sessions.

Managing the evolution

The NSF planning grant was an intense and carefully managed program and probably had a greater effect on public education in Montgomery County than any other single event in the county's history. It came as a *deus ex machina*, a rather unexpected event that galvanized interest and concern for education throughout the community. It was accompanied by much publicity, discussion, experimentation, demonstration, and learning. Its effects penetrated the political arena as well—the county held its first school board elections, changing the board significantly in the process. Previously, members of the school board had been appointed by the County Board of Supervisors. Although a number of circumstances magnified the effect of the NSF grant, the process arguably was fueled by the accessibility afforded by the BEV in Montgomery County as well as the astonishing growth of electronic communication and networked resources including e-mail and the World Wide Web. Managing the evolution of a virtual school is a brand new experience requiring network connections, educational reform, networked information competence, leadership, and financial backup.

Dramatic as those events have been, the schools and their university counterparts had much to learn about each other and the business of networking and education reform. Firsthand experience with managing networked education taught them more than anything else. They learned by doing and from mistakes each step of the way. Technology, new roles, budget, and scheduling presented new challenges. For example, the few yards separating Riner Elementary School from its T1 connection turned out to require a ditch. That had to be bid, whereupon the selected contractor declared bankruptcy. Almost everything was over-budget.

Next came the problems of actually cabling the classrooms and installing network cards and software on the barely adequate collection of computers in the classrooms, all this with a minimal and badly over-worked staff. With so many parties involved, the network connectivity did not materialize on schedule either. An early dose of reality for the BEV team (faculty and staff of Virginia Tech) came when they experienced firsthand how difficult it is to organize a physically dispersed group of K–12 school faculty and administrators working on schedules that make them generally inaccessible during the day. For the most part, the teachers were working on their own time, and the project team quickly recognized the need to support work at home with equipment and flexible scheduling.

The plans for the future use of computer networking to change the quality and effectiveness of education require changes in teaching style, adoption of new models of education, and competency in networked resources. First and foremost, the mere presence in the classroom of a networked computer has little effect on the educational process and outcome. It is well argued that the use of technology does not by itself improve learning; rather, it must be integrated with reforms in teaching styles and strategies like those under the NSF-supported Statewide Systemic Initiative (SSI) (e.g., making instruction active and student-centered; developing critical thinking and other higher order problem-solving skills; providing the avenue for students to link mathematics, science, and technology with authentic or real-life applications; and communicating high expectations for all students).

For the Montogomery School System, reform requires the adoption of learner-centered [6] or constructivist models of education [7] and effective use of Internet resources. For most teachers, effective use of the Internet is an acquired skill. The Net is an anarchic place; it gives up its secrets only to those who approach it with a methodology, patience, determination, and a sense of adventure. Only the most creative and accomplished teachers have the ability to map the peculiar resources of the Internet into a directed series of lessons that achieve their educational objectives.

The need to transform pedagogy, to learn new technology, and to adjust to the challenges of teaching in a networked environment is but the beginning of the human problem for the teachers. Having taught with substantially the same style for many years, most teachers are not accustomed to an environment in which they must learn faster than their students and on their own time. Developing the necessary training and support structure is costly, and providing the motivation and rewards that a teacher will find compelling is still more difficult. After all, teachers want to focus on their educational agendas, and many view the need to deal with ASCII codes, broken cables, GIF files, and unusable freeware as an intrusion or a diversion of their energies. Finally, when the day comes for the delivery of the first network-based lesson, a teacher realizes that 20 little bodies do not fit easily around a 14-inch display and that some students in the back may start shooting spitwads. Indeed, networking technology can be fragile. Files can vanish, computers can break, and entire servers can go down for weeks at a time, usually when they are needed the most. Learning about class management and coping with scarce or unreliable resources are other aspects of classroom reform. Network-based instruction requires extra planning and preparation, and teachers need to consider in advance the possibility of technical failures.

Despite its various problems, the Internet is a natural ally of education reform and constructivist models of learning, which have a direct impact on school children. It can (1) make instruction active and student centered; (2) help develop critical thinking and other higher-order problem-solving skills; (3) provide the avenue for students to link mathematics, science, and technology with authentic or real-life applications; and (4) communicate high expectations for all students. American children are accustomed to stimulating audiovisual media environments and feel comfortable with the one provided by networking. Their ease with interactive media empowers them to utilize network-based learning, which is also learner-directed, responsive, and based on an unimaginably large global information base that grows ever larger.

Electronic communication offers rich collaborative applications and has the potential to generate a profound impact on education as well as the general community. Computer networks provide a communication medium that has never existed before for group participation, collaboration, and interaction [8]. Recent research shows that in the classroom, online interaction is more than supplemental [9, 10]. Computer-mediated communication affects social and psychological aspects of group communication, as well as the quality and the extent of discussion, in several specific ways:

- It increases the overall level of discussion of the entire group.

- It increases student participation, especially from those who do not participate in face-to-face settings.

- It is sexually neutral. Some studies (Ruberg, 1994) show that women participate as often as men and are as often early initiators as men in an electronic discusion group.

- It increases understanding of material.

- It increases morale and motivation.

- It increases community interaction and involvement in the educational process, thereby improving the educational experience and facilitating long-term sustainability of network resources in the schools.

The transition to effective learning in a networked environment poses a challenge to those setting standards of learning. The Commonwealth of Virginia has established a set of standards of learning (Figure 8.7), which itself entails disparaties between expectations and reality. The standards specify what a student should master in each school

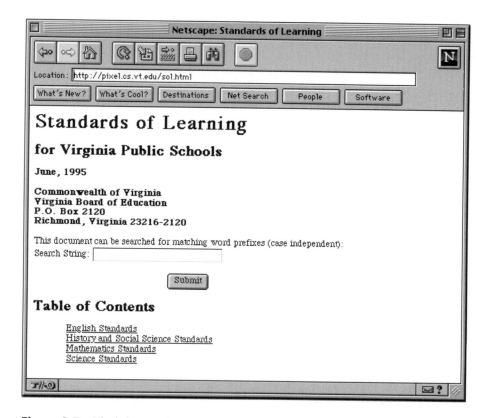

Figure 8.7   Virginia standards of learning. ©1997 Netscape Communications Corporation.

year (see http://pixel.cs.vt.edu/sol.html). Adoption of the standards is encouraged in part by new state funds to support the purchase of one computer for each classroom, or about $900,000 in the case of the Montgomery County Public Schools. The 1996 document sets aggressive standards for achievement in computing and networking for the fifth and eighth school years. Students are required to know how computers work, and they must have a working knowledge of information processing and retrieval, desktop publishing, and networking, including network publishing. However, there is a lack of standards regarding methodology, testing, and and evaluation of learning. It is left to the school systems to determine how that information is to be taught. Difficulty in achieving the goals in the standards is complicated by the fact that often classes are taught by teachers who themselves have had little or no exposure to the content specified in the standards of learning.

The fundamental challenge boils down to who will manage the evolution of a virtual school. The realities of the situation call for leadership and energy, vision and wisdom, power and financial authority. A community must either build a coalition of partners who agree on the goals or be fortunate to have in place leaders with those qualities. State or national mandates such as Virginia's standards of learning will also help. We have observed that major reform efforts are much more effective when driven by a single individual or a small group. In Montgomery County, communities are responding, but these responses are insufficient to drive the massive changes that are taking place there. Instead, change was initiated through the conceptual leadership provided by the BEV and supported initially by the NSF grant. Within the schools, the administration has been supportive, but leadership has come from a small group of master teachers who have the vision for major reform. Such a bottom-up approach can be problematic because of the difficulty of convincing disparate groups to work as a team. For example, a network-based reform effort in the schools requires a close working relationship between the technology and support staff and those responsible for curriculum.

Managing the evolution of the virtual school requires leadership in managing the funding process. Because the community has the ultimate authority to authorize funds for network-based reform, the schools must make an exceptional effort to work with community leaders to formulate and promote a clear case for financial support. That calls for tireless and articulate leadership in managing the political process. What is relatively new to the Montgomery County schools is the competition for large-scale federal, state, and corporate funds. All the large grants have so far come in partnership with the BEV and Virginia Tech. The BEV and Virginia Tech have helped the schools to understand the funding process and establish an infrastructure for proposal preparation and processing. As a result, the schools hired a dedicated grant writer in 1995, on the recommendation of the technology committee. Currently, the school board is dealing with the mechanics of the approval process and the ever more common requirement for the schools to cost-share significant portions of these projects with very tight public funds. The problem of funding school technology will be difficult everywhere. The Fairfax County,

Virginia school systems, generally regarded as one of the most affluent in the nation, recently proposed that parents purchase laptops for their students, beginning in the sixth grade. Even those schools have had difficulty financing equipment as well as the extensive training required to enable teachers to make the necessary changes in their classrooms.

Integrating networking into the community and the schools

Most users' experience with networks over the past 10 years or so has been in one or two contexts: the largely anonymous, geographically diverse Internet community and the usually hierarchical and autocratic local organization (a university, a corporate employer) that has provided the local connection. As part of a homogeneous organization, use has been dictated by policy and protocol with little opportunity (or need, perhaps) for the kind of continuous needs-education cycle that has been vital to promoting the use of the network in the larger BEV community, which includes the schools.

Because the BEV is technology-intensive, many teachers and community residents feel suspicious, intimidated, or even disenfranchised by it, regardless of their social or economic backgrounds. Others, like many Europeans with strong traditions of personal socialization, simply fail to see that computerization and networking adds value to their lives. Without a tradition of involvement in the information revolution, parents, teachers, and school administrators may not understand the dramatic potential of computer networking for improving or reforming both the curriculum and the process of education. What actually happens in a particular community depends on which of these groups becomes the network-literate technology adopter. Regardless of which community segment drives reform, we find that the children themselves become instant converts, since networking provides a new means to satisfy their extraordinary curiosity. In fact, many parents and teachers experience a role inversion, in which their children and students learn and practice skills and techniques they themselves have not yet mastered.

The community education problem has been difficult and challenging. Networking often must be explained and demonstrated time and again before users are convinced of its value in their daily affairs. Even a young child can learn to put a quarter into a vending machine to obtain a newspaper. In reality, surfing the Web at the public library is not much more difficult, but many who have not done so believe that it is. That sets apart county socioeconomic groups and creates problems that can be solved only by education. Education has two aspects: (1) specific training to integrate network resources and tools into daily activities and (2) public education in general. As we will see, those issues are not separate but rather part of a deep social realignment that is taking place in Montgomery County. Education is being attacked on numerous fronts: through the educational institutions, through public facilities such as the libraries, through social organizations, and even through educational materials provided on the network that caused the problem in the first place. As with other major social revolutions in recent history, networked computing may become so ubiquitous that people will no longer think about it.

In a somewhat counterintuitive finding, people in the Blacksburg community are not particularly impressed with the vast resources of the Internet [11]. What seems to interest people most is better access to local, real, substantial services, rather than to the nearly infinite reaches of cyberspace. But even that enthusiasm triggers a massive education effort: Merchants and businesspeople must be taught how to package, sell, and service their merchandise on the network. Local governments must be educated about networking and why it may be a good thing to give citizens better and more frequent access to government information. Local civic, volunteer, and social groups can leverage the tools of the Net to good effect, but the often casual organizational nature of such groups makes educating them difficult.

Network authoring on the Web is another acquired art, a combination of communication, psychology, marketing, and design skills. While the syntax of the underlying language, HTML, is easily learned, it takes a long time for designers to understand how to use it well for effective communication. A Web document is a human-computer interface to the information world, and all the understanding of the design process that we have acquired in the past two decades needs to be applied to those documents. Every Web document has a reason for its existence. More often than not, the authors are promoting ideas, explaining, cajoling, selling, or promoting their credibility. The children are learning about this, too, learning that part of life involves expression and making others aware of one's existence.

One of the aspects of the work in the educational arena that continues to be astounding is how little most community members know and understand about their public schools. Parents send their children to the busses every morning, deal with their homework at night, and perhaps attend an occasional PTA or teacher meeting, but otherwise they have little insight into the educational process and how it is supported by their taxes. This is peculiar, because freedom of information is fundamental to the democratic process, and an informed community should be fundamental to the conduct of its affairs. There seem to be several explanations. For one thing, the complexity of governmental affairs is so great that most people could not absorb all the information even if they could find it. Perhaps more important, the information tends to be available but unpublished, and there is often little motivation to publish it. In the first place, traditional publication is expensive, and in some cases, government officials may choose to withhold information in order to exert a measure of control.

Since the coming of the BEV, there have been significant changes in the accessibility of information, and that is beginning to affect public decision-making. One can obtain the names of the teachers in a particular school, write the board of supervisors, read the proposed county budget, peruse the state standards of learning document for the schools or some controversial legislation, or fire off an angry letter to the school superintendent. Despite the tendencies of people not to read, there is evidence that the members of the Blacksburg community are becoming better informed. At least one school board member updates his constituency regularly through an e-mailing list. How that will play out in community decision making and in the evolution of public institutions such as the schools is yet to be determined.

The uncontrolled nature of network-based information has created a new concern about the many documents on the Internet that are not suitable for K–12 students. Those risks have received so much attention in the public media that parents and teachers who do not understand the technology often overreact. In Montgomery County, as in many other school districts, network policy combines student supervision and direction with calm, matter-of-fact discussions with parents. The principal sends home with each student a letter with an acceptable use policy form for signature by both parent and child. Parents are made aware not only of the risks but also of the significant benefits to their children. So far, that rational approach seems to have prevented major problems.

At least two major obstacles must be confronted before significant reform takes place in the educational system. One obstacle is financial, and the second has come to be called "technology refusal." The financial obstacle concerns technology acquisition, which is the basis of network-based reform. Technology refusal deals with the actual changes that need to take place in the educational process. For the sake of argument, let us make some simple estimates of what is required to acquire the technology necessary for educational reform. At $2,000 per computer and a ratio of one computer to four students and faculty members, equipping the public schools in Montgomery County would require an investment of $5 million. One month (20 days) of training per staff member would add $660,000 to that total in release time alone, and many doubt that that training would be enough to significantly change the classroom process for most teachers. Then there are the continuing costs of network connections, hardware and software upgrades, and support personnel, which could easily add $1 million per year. In a society as wealthy as ours, that goal is easily achievable, provided there is a unity and resolve in the community that the expenditure is necessary for the future of our society. In fact, a one-year initiative to acquire the required facilities would add only 20% to the county tax bill for that year.

Other investigators [12] have studied problems associated with technological change in school organizations and have concluded that because of the complexity of the system and of constraints that must be satisfied by a school system and its staff, real educational change associated with the introduction of new technology is difficult to achieve. Technology introduces values and practices that do not always integrate smoothly with those of the school organization. Hodas attributes the general lack of impact of new technology on teachers, administrators, and students not to implementation failures but rather to a large disparity of the values embedded in the technology from those of the school organizations. The organizational momentum does not support the changes associated with effective deployment of new technologies, with the result that little reform is achieved.

In the Montgomery County schools, the project implementors are attacking both problems but, sadly, are far from the $5-million facilities goal. Through the NTIA/TIIAP grant from the Department of Commerce, there is now a fully equipped training laboratory in a single school, and a county-wide network technology training program has begun. So far only a few teachers have fundamentally changed the way they teach; it remains to be seen, as the facilities evolve, whether pedagogical reform will actually occur across the school system. Even training has been proceeding slowly; many teachers know that back in the classroom they do not yet have the facilities they need.

A conclusion reached by several of those involved with the NSF planning grant is that, to introduce educational networking properly, it will be necessary to create a separate school, perhaps a so-called exemplary school. That school would adopt the best possible curriculum, the best teachers, and the best technology and then propagate the new culture outward into the entire school system. Some observers believe that a curriculum such as an international baccalaureate curriculum seems consistent with networking technology, and that is now a topic of serious discussion. Educational change in the schools will be difficult without corresponding changes within families and the community. One of the best potential agents of change in the community is the electronic village itself, but it will take massive education to turn it into an effective agent of change. Ways to draw the parents into closer collaborative relationships with the schools and with government are being explored. If the project is successful, people will no longer have to move physically to acquire information and to learn. Information will come to those who seek it online.

Usability concerns

One does not have to spend a great deal of time with teachers to understand the incredible sensitivity of their attitudes toward computing to the human factors of their environment. *Human-computer interface* (HCI) professionals have long been surprised at how important, long-range decisions are being made without adequate concern for the human factors of the school computing environment. That environment consists of specific program and network clients and their particular interfaces and also of the computer itself, its peripherals, the network connections, the furniture, the lighting, the displays, the technical reliability, and all that is required for quality classroom instruction to take place. The usability of the environment refers to the ease with which work can actually be accomplished in the environment, as well as to the accuracy of the work, the ease with which the new technology can be learned, and the satisfaction that the users perceive with their facilities. Unless the usability of all components is carefully considered, the ability of the teacher to conduct a class is easily compromised. It is counterproductive to invest heavily in classroom technology, only to have it fail because of poor usability. Frequently when the technology fails, it is difficult to convince a teacher to try again.

Some examples illustrate the importance of usability. Consider the importance of a sensible computing configuration. Because of shortages of funds, computers often are ordered with minimal amounts of memory and storage so that equipment can be purchased for more teachers. If it is subsequently determined that a critical application cannot be run, the equipment will not be used, and the financial investment will not produce the desired results. Because changes in computing occur so rapidly, saving a few dollars on a configuration may substantially accelerate obsolescence and thereby reduce the value of the investment.

Inadequate program-user interfaces may also decrease a teacher's capabilities. The Internet, for example, is replete with different FTP clients with a variety of interfaces that range from textual to graphic. Some interfaces that are wonderfully graphic may crash repeatedly or have no recovery mechanism for frequent server timeouts. How can substantive decisions be made about the creation of networking environments in the schools without those environments first being optimized and their usability tested? In general, because of their dependence on scarce public funds, teachers have to put up with human factors that home-computer users would not tolerate.

Much interesting and potentially useful software is being produced and made available without adequate usability testing. In 1995, the authors performed a formative evaluation of the popular videoconferencing program from Cornell University, called CU-SeeMe, which we found conceptually fascinating. Multiparty videoconferencing poses some interesting usability problems that suggest themselves quickly when one observes a typical group of new users testing the software, tapping on the microphone, and typing, "Did you hear me?" Visually, the numerous faces on the display screen suggest the gorilla's view of the zoo. One problem with the version we tested was that the visual cues were confused. For example, a speaker addressing one party may seem to be moving her lips because of the low frame rate, and the darkening speaker icon may be missed because of its position on the display. On the other hand, another participant may think that he has been addressed instead. Generally, the participants spend their time scanning the video windows and associated interfaces for activity cues, to the extent that they have difficulty participating in the conference itself.

This formative evaluation of videoconferencing shareware was actually addressed at novice users in a conference with two or three participants. The results suggested that users at all computing expertise levels had difficulty getting started with the interface because of a number of design features. Although one subject walked out of the experiment, the rest seemed to learn the interface quickly because it is not that extensive. The authors performed this evaluation as the basis for a formal study of the interface's planned use in fifth grade geography instruction. In that study, the predictions from our formative studies were confirmed in a pretest, and usability problems created significant difficulties in initial trials. However, the fast learning curve for the software plus attention to some of the details of the classroom setup solved most of the problems in subsequent sessions.

Assessment and prospects

Those associated with the BEV failed to anticipate the massive education effort that would be required to give the public technical and social facility with networking. As a result of many hours of training and education at the university, the public library, the schools, a friend's place, or elsewhere, participants are learning new ways to conduct daily affairs. Information and service providers are responding in kind with new and innovative offerings. People can more easily obtain information about individual schools in Montgomery County through their extensive Web pages and can play a more active role in problem solving in the schools through online discussions with district representatives, administrators, and teachers.

Schools must allocate time for teachers to explore network resources, to reflect, and to synthesize new approaches to their educational program. They must provide technical assistance at a personal level to help teachers overcome both technical and conceptual barriers. Teachers generally are problem-oriented and put off by the necessity of dealing with technological details that seem irrelevant to their goals. They need mentoring and encouragement from colleagues who have already mastered new skills. The negative human factors that seem to characterize school computing discourage teachers from serious involvement with network-based instruction. Computers must boot when turned on, connections must be quick and adequate, and there must be support personnel to tend immediately to problems. A malfunctioning mouse must not be allowed to sabotage a week's carefully laid lesson plans.

The Internet services available in the popular networking suite distributed by the BEV make substantial contributions to private mail communications, published mail communications, and digital libraries. Recently input components have been added to Web HTML scripts to permit users to return information to a server from Web browsers. Still, the basic Web and Gopher paradigm is an asynchronous one in which server processes vanish after transmitting a page to a client browser. In education, active learning is enhanced through processes of collaboration in which learners collaborate with software agents, with each other, and with groups, possibly solving complex graphical problems through a shared workspace and world model. Although there are several possible mechanisms by which collaboration (which is known as synchronous collaboration) could be supported by the Web, little work to date supports education aside from a few feasibility studies performed at networking laboratories around the country. It is the authors' belief that network-based collaboration is a fundamental and necessary addition to the educational networking suite.

As schools and communities gain greater access to information networks, they will use those networks to link schools and families closer together in support of the teaching and learning process. However, along with the potential for greater linkage comes the potential for increased disparity for children who live in homes without the economic means to provide network access. Today, only one-third of American households have personal computers, and one estimate suggests that most of the households that can afford a computer already have one [13]. A more recent study is more optimistic, noting much activity by first-time computer buyers and the possibility that declining prices may spur computer sales among lower income households [14]. Still, if the schools do not act to provide adequate computing support for their charges, then a serious educational disparity is a certainty. The schools must find creative ways to enfranchise lower socioeconomic families and other have-nots in the community. One of our goals is to increase the availability of computing facilities in the county libraries and to make new school computing laboratories open to the community after school hours.

Resources for education, even more so than in most other sectors, have become increasingly scarce. So that human and capital resources are allocated optimally for education, we will continue to seek and to disseminate the best evidence about the circumstances under which innovations in teaching strategies and resources achieve greatest effectiveness. A nagging problem for educators as well as for community members and their representatives is that to date there is little evidence that constructivist network-based education will translate into student achievement. There is not even a consensus on how that might be measured. Teachers who have adopted network-based learner-centered teaching styles know how well those styles work in the classroom, but no longitudinal studies have yet been completed that demonstrate the changes in educational outcomes. However, with estimates that "fully 60% of jobs in 2000 will require a working knowledge of information technologies" [15], national educational reform is no longer an option but a mandate.

Whether administrators in the thousands of independent educational systems across the country have the wisdom and the vision to respond—or even the ability to comprehend what has happened in information technology since 1993—is an open question. With less and less centralized direction, increased disparities among school systems nationally is a real possibility. Although state and federal governments are providing funding and incentives for network connectivity, the actual responsibility for educational reform is being left to the individual communities.

As the 1995–96 school year draws to a close, 150 of the 602 school classrooms in the Montgomery County public schools have an Ethernet connection, though not necessarily a networked computer. It is hoped that by the end of the next year most classrooms will have both Internet connectivity and a networked computer. For those who had hoped that the schools would receive increased funding in 1996–97, 1996 was catastrophic, with a total local budget increase of only $70,169. The technology program will continue with state funds, but it is notable that since 1990, when state and local contributions to K–12 education were roughly equal, the local contribution has declined steadily, to less than 42%. There is sharp division on the county board of supervisors over the school budget; it will be extremely interesting to see whether the sociological changes that have been taking place in Montgomery County will have their ultimate expression at the ballot box in the next election.

References

[1] Labonski, R., *"Preliminary Assessment of NCCN Region Demographics,"* Consultant's report presented to New Century Communications Network, Inc., Richard Labonski Associates, P.C., Silver Spring, MD, 1994.

[2] Arrington, L., *"Technology Plan for Montgomery County Public School System,"* Montgomery County School Administration, Christiansburg, VA, 1994.

[3] Latham, G., and J. Burnham, "Innovative Methods for Serving Rural Handicapped Children," *School Psychology*, 1985, 14(10), pp. 438–443.

[4] Bull, G., I. Hill, K. Guyre, and T. Sigmon, "Building an Electronic Academic Village: Virginia's Public Education Network," *Educational Technology*, April 1991, pp. 30–33.

[5] Cohen, M., and N. Miyake, "A Worldwide Intercultural Network: Exploring Electronic Messaging for Instruction," *Instructional Science*, 1986, 15(3), pp. 257–273.

[6] Norman, D. A., and J. C. Spohrer, "Learner Centered Education," *IEEE Computer*, 1996, 39(4), pp. 24–27.

[7] Ehrmann, S., "Asking the Right Questions: What Does Research Tell Us About Technology and Higher Learning?" *Change*, March/April 1995, pp. 20–27.

[8] Neuman, R., *The Future of the Mass Audience*, New York: Cambridge University Press, 1991.

[9] Sproull, L., and S. Kiesler, *Connections: New Ways of Working in the Networked Organization*, Cambridge, MA: MIT Press, 1991.

[10] Ruberg, L., *"Computer-Mediated Communication Environment,"* Virginia Polytechnic Institute and State University, unpublished Ph.D. dissertation, 1994.

[11] Patterson, S., and A. Kavanaugh, "Rural Users' Expectations of the Information Superhighway," *Media Information Australia*, 74, 1994, pp. 57–61.

[12] Hodas, S., "Technology Refusal and the Organizational Culture of Schools," *Educational Policy Analysis Archives*, 1993, 1(10). Available through the listserver edployar@asuvm.inre.asu.edu.

[13] "Computing the Real Costs," *Montgomery County News Messenger*, April 3, 1996, p. A4.

[14] Hill, G. C., "Tally of Homes With PCs Increased 16% Last Year," *Wall Street Journal*, May 21, 1996.

[15] U.S. National Information Infrastructure Advisory Council, "Realizing the Benefits," *KickStart Initiative, Connecting America's Communities to the Information Superhighway*, 1996.

9

Learning and Teaching in a Virtual School

by Melissa N. Matusevich

I MAGINE AN elementary school classroom where students converse by e-mail with William Shakespeare and Mark Twain. In the same classroom, students compare geographic information with a faraway school via video teleconferencing. Students use a computer to interactively talk with a graduate student in Wales and learn that his world is five hours ahead of theirs. Imagine, too, that as a comet strikes Jupiter or a volcano erupts in New Zealand, the students in this class view those events as they occur. A classroom that exists only in the future, you say? In Blacksburg, Virginia, with the school connected by a T1 line to the Internet via the BEV, these classroom activities are common occurrences.

Technology and education

Information technology can and should be used to enhance instruction because it provides teachers with one more effective tool. Technology has always affected education: The printing press allowed textbooks to be developed, and the replacement of slates and chalk with pencil and paper permitted students' writing to be preserved. In the late 1950s and 1960s, television was utilized as a means of teaching large groups of students, albeit ineffectively. Today, a new wave of technology is beginning to cause repercussions in schools that will forever change how students are taught. That technology is the personal computer and an Internet connection.

Education is being partially transformed by new technologies. At one time, students could learn a small but fixed body of knowledge. Today, the enormous amount of available information, coupled with the fact that the existing knowledge in the world continues to double at an ever increasing rate, requires a transformative approach to education. Students of today must learn how to evaluate, manage, and use information—not just regurgitate it.

In a technology-rich environment, one must remember that the educational focus is on learning and instructional goals rather than the technology itself. Technology provides tools or vehicles for delivering instruction. How technology is used is of utmost importance; an electronic worksheet is still a worksheet. Most worksheets focus on the acquisition of low-level skills (such as alphabetizing lists of words or capitalizing proper nouns), a virtually impossible task when teachers utilize the Internet to enhance student learning.

Studies show the many observable changes in technology-rich classrooms:

- There is a shift from whole-class to small-group instruction.

- Coaching, rather than lecture and recitation, occurs.

- Students are more actively engaged.

- Students become more cooperative and less competitive.

- Students or groups of students learn different things instead of all students learning the same thing.

- There is an integration of both visual and verbal thinking instead of the primacy of verbal thinking [1].

- Student self-esteem is strong.

- Student attendance improves, and discipline problems are reduced.

- Students come in on their own time and often give up recess to work on projects [2].

Soon after obtaining my first e-mail account and Internet access in 1991, I sensed that a veritable gold mine was literally at my fingertips, and I was right. I began using Internet resources daily with my students by posting questions, for which we could locate no answers, to appropriate newsgroups and to exchange e-mail with other users in locations both inside and outside the United States. All beginning attempts focused on merely the gathering of information. It was not until later that I realized that having my students analyze information and publish their results or other work were important. The Internet is one more tool for a teacher to use and can provide a powerful set of resources. Here are a few reasons why I believe teachers should use Internet resources in their classrooms:

- Any question a student may have for which local resources can provide no answer can be posted to a Usenet newsgroup. Generally, an expert in the field responds. School libraries are far too limited, and students often become frustrated when they are unable to locate the information they are seeking.

- Students can access information while it is current. When a major world event occurs, such as a comet streaking across the sky in the southern hemisphere, students can learn about it immediately. Most textbooks are years behind, expensive, and out-of-date soon after they are published, and many interesting yet ephemeral events never make it into textbooks.

- Students can converse with people worldwide instantly using e-mail, interactive talk, and video teleconferencing. I have had students send and receive a reply to their e-mail query to Norway in 10 minutes.

- Students need to learn how to access information. No one can possibly memorize everything he or she needs to know. Knowledge continues to double at a fast pace. It is important for students to learn to use information technology to their advantage. Students must be prepared not only to deal with the body of knowledge that currently exists but also to discover the body of knowledge yet to exist.

- Knowledge has a shelf life. How many of us have made students memorize facts about subjects such as the solar system that are no longer true? I shudder when I think of all the incorrect information I required students to memorize 20 years ago.

Success in any career hinges on the ability to communicate effectively both inside and outside the organization in which one is employed. Having good ideas is not enough; one must also have a facility for sharing those ideas with others. The Internet is primarily a medium for communication on a global scale. Global interaction has become a way of life in business and industry as well as in other fields.

Constructivist theory

Many theories of learning have been proposed in the last 100 years. One of these theories, constructivism, seems highly relevant to these times, in which modern technology is significantly affecting society in our daily lives. Public schools have and will continue to reflect societal change. There is a strong link between effective use of modern technology and the theory of constructivism.

Constructivism is the belief that one constructs knowledge from one's experiences and mental structures, which are used to interpret objects and events [3]. My view of the external world differs from your view because of my unique set of experiences. Peter Senge put it this way: "We don't describe the world we see; we see the world we can describe" [4].

An important component of constructivist theory is the focus of a child's education on authentic tasks. Those tasks have real-world relevance and are integrated across the curriculum. Because it is impossible for all students to become masters of all content areas, instruction is grounded in a meaningful context. According to constructivist theory, children learn, not incrementally, but from the whole to the part. The ideas and the interests of children drive the learning process, with flexible teachers sometimes being the givers of knowledge and at other times the facilitators. In a constructivist setting, the clock does not rule, because dynamic learning is the constant and time the variable.

Juxtaposed to constructivism as a model for teaching and learning is the didactic model in which teachers "cover" material and students are taught to memorize content rather than to seek deeper meaning. Generally, the teacher lectures, the passive student takes notes, and an objective measure is given to test knowledge. This teaching method is widely used because it is how many teachers were trained to teach. Most teachers were taught that way when they were in public school and had didactic teaching modeled over their entire public school experience. Thus, even though students come to us from a technology-enriched environment where they control information flow, they are most often expected to fit into an educational institution unchanged by the technology that is sweeping society.

In the traditional incremental, didactic model, the student is fed a succession of facts, one after the other, at the end of which, the student is expected to integrate those facts into a coherent whole. Constructivism takes the opposite view; its whole-to-part approach to teaching implies that learning is nonlinear. Discoveries in one area may lead students to pursue tangential avenues appropriate to the topic under investigation. Embodiments of this nonlinear approach to presenting information are hypertext and hypermedia, as popularized by the WWW and desktop multimedia computing.

Because of the resistance of educators to abandon didactic methodology, the integration of technology into the classroom has often been rejected. Trying to fit new technologies into this old paradigm creates a situation in which teachers refuse to pursue using technology even when it is provided to them [5]. In instances in which technology-based systems have been used to teach mechanical skills, the result is that the technology is used for low-level instruction rather than for the rich instruction that can lead to higher level thinking. Skill-based computer games that focus on teaching children to memorize math facts or selecting the correct part of speech from a list of words became quite popular in the early 1980s. No high-priced piece of technology is required for this type of teaching; a cheap worksheet can do the job just as efficiently. In other words, automating didactic instruction will have little if any effect on improving educational outcomes. The learning of superficial information via high-powered technology still results in the learning of superficial information. Technology, however, has much more to offer the learner. It opens up new avenues and creates environments in which learning, rather than memorization, takes place.

For technology to be used effectively in classrooms, systemic change must occur. As technology becomes a part of our everyday lives, and because schools reflect societal changes, the integration of technology into school settings may force the use of constructivist methodology. As computers become an integral part of everyday life, it is possible that teachers will no longer have a choice but will be compelled to use a constructivist approach in a technology-rich environment. In other words, constructivism might be viewed as a theory waiting for technology to catch up.

Constructivism is a natural fit with technology. Why?—because computers undermine the didactic, lecture methodology and instead promote students as self-directed learners. Self-directed learners become lifelong learners because they have the skills to continue to direct their own learning and are not dependent on knowledge fed to them. Rather, they know how to acquire knowledge themselves. As an old proverb aptly states, "Give a man a fish, and you feed him for a day. Teach a man how to fish, and you feed him for a lifetime."

The use of new technologies in educational settings allows students to become empowered by gaining access to real data and working on authentic problems. One example that highlights this is the use of a standard print encyclopedia versus a CD-ROM encyclopedia. When seeking information on a topic, few children will explore anything but the first print encyclopedia pages he or she locates, ignoring cross-references. Using a CD-ROM encyclopedia, which is structured nonlinearly, the student easily moves from link to link, often uncovering more specific information not found at the initial search location. Even the physical effort required to follow a cross-reference in a standard print encyclopedia, which may involve retrieving separate volumes, makes such a process tedious, unwieldy, and unattractive to the student. Children become eager to share what they have found. I learned much about sharks this year because of the information a student located on such a search. Brian, a seventh grader, sought information first from a CD-ROM encyclopedia and then from the World Wide Web through the school's BEV connection. He became so excited about what he found that he could not wait to share it with me. He even located a quiz about sharks, which he asked me to take. This reversal of roles, as teacher becomes learner, is not uncommon in technology-rich classrooms.

We are in an information age. There is an increasing shift in the work place toward critical information evaluation skills. Barr says, "If we wish to prepare students for lifelong learning, we must begin to introduce them to the tools which they will use in the careers they pursue after their formal education is completed" [6].

There continues, however, to be much resistance to constructivism, and it is not widely accepted in the public school arena. The general public is often suspicious of educational practices that differ from what they experienced. They feel far more comfortable with the textbook approach, which they believed worked well for them. This is like driving a 1964 model car with none of the features we take for granted today. While the car may get you where you want to go, it is not necessarily the best way to travel. Many statewide assessment programs are not in alignment with constructivist theory. Students are often required to perform on standardized tests that do not assess what they are learning. Also, the structure of the school day causes most classes to be fragmented into small blocks of time.

Lack of funding and no clear vision keep systemic change from occurring as rapidly as technology is evolving. Too many teachers are unaware of much of what is available for them in the world of technology. For a major paradigm shift to occur, much groundwork must be laid. Besides the acquisition of appropriate technology, schools need time and support to become comfortable with it. Teachers need training in constructivist theory and its application to the world of information technology and the classroom. (The public must be converted as well.) Ways to effectively integrate technology into the required curriculum must be explored. All too often, first-year teachers enter a classroom having never turned on a computer! Having worked with many student teachers for the past

20 years, I have been surprised to find that most professors in colleges of education do not use computers themselves and, therefore, do not require their students to learn to use them either. Comprehensive training in the integration of technology into the public school classroom is sorely needed, and it appears to be lacking from current teacher training. As a result, new teachers are often "more like their predecessors who graduated decades earlier than they are like today's children" [7]. The result, according to Strommen and Lincoln, is that there exists an estrangement of schools from both society in general and from the children who attend those schools [7].

Given that technology has a valuable place in the classroom, the question now becomes not *why* to use it but *how* to use it. Any teacher, with even modest resources, can transform a classroom into a dynamic learning environment where students are active learners using higher level cognitive skills. By integrating information technologies into the required curriculum, a teacher can create a rich environment in which students pursue solutions to real-world problems.

Constructivist practice: BEV + MCPS = constructivist classroom

I began teaching in Montgomery County public schools (MCPS) in the early 1970s, spending my time as a teacher of fourth and fifth grades, with a three-year stint as a staff member of the gifted-student program during the late 1980s. During these many years, I endeavored to advance my knowledge of learning theory and classroom practice in efforts to meet more effectively the needs of my students. Always open to new ideas, I was somewhat a pioneer in implementing new methodologies that were firmly grounded in research. Best practice was my goal.

In 1991, when I acquired my first e-mail account, I was teaching fifth grade at Margaret Beeks Elementary School. This was two years before BEV accounts became available. Our statewide system, VAPEN, gave a free computer account to any teacher requesting one. VAPEN was a text-only system. At that time, I had at my disposal an Apple IIe computer, a 2,400-baud modem, and a telephone line that I shared with six other teachers. Within this low-tech world, my students and I made many discoveries about teaching and learning that forever changed the way I view education. Any teacher, with even minimal resources, can do the same, as we shall see.

My first attempts at using the Internet included a focus on reading and responding to Usenet newsgroups. Usenet is like a global bulletin board. It is organized as a hierarchy of topics; for example, sci.math is the name of a newsgroup devoted to mathematics. The prefix "sci" denotes that the newsgroup is part of the hierarchy "science." The hierarchy can be many levels deep. For example, comp.unix.admin is a newsgroup devoted to the topic of administering Unix computer systems. People from all around the world post messages and dialogue about pertinent topics. The messages are similar to e-mail but are posted to a newsgroup for all to read rather than being sent to an individual. I subscribed to a few scholarly newsgroups such as sci.astro and sci.math so I could learn what the world of Internet newsgroups was all about. (VAPEN limited the newsgroups to which we teachers could subscribe; almost all the groups were curriculum-related.) The discoveries I made heartened me. I noticed that the majority of posters were located at major universities worldwide. I also realized that real-world questions were being asked and that responses were given by others in the field. I noticed, too, that international e-mail addresses ended in two-letter codes, and it was not long before I made the connection between the two-letter code and the country of origin. Thus, an e-mail address that ended in "uk" meant that the post was from a site somewhere in the United Kingdom.

It is often thought that a teacher with a large class and one computer has little opportunity for integrating technology into the curriculum. The following stories illustrate how wrong that assumption is.

One of my students asked a question about a tuatara, an animal native to Australia, for which we could find no satisfactory answer locally. On a hunch, I decided to post the student's query to the sci.bio newsgroup. I expressed concern that I had a 10-year-old student who had spent much time seeking information with poor results and who was growing frustrated with the researching techniques available to him. I posed his question, and it was not long before many biologists responded by e-mail with an abundance of information that excited my entire class. A lively exchange ensued between one of the biologists and my student, who was motivated to extend his research on the tuatara. As a result of this success, we began to view the Internet as a potentially invaluable source of high-quality information. We started utilizing the medium for the posting of questions, but only after local resources had been exhausted. No one wants to do students' work for them but will gladly come to the students' aid after they have carefully delineated the efforts to which they have gone in seeking out information. To this date, no question a student has posted has remained unanswered. One side benefit is that students often receive more information than they can utilize, which requires that students learn a new skill: how to assimilate information and to decide what is relevant to their needs.

It was using this method that a fourth-grade class I was teaching stumbled into writing to Mark Twain himself! A group of students had read *Tom Sawyer Abroad*, a short novel by Twain that gave a flavor of his style of writing. The book ends abruptly with a most unsatisfying conclusion. One student was so upset by this that she asked, "How can Mark Twain be considered a good writer? This ending is awful!" I had no answer, so we did what we always do. We posted a message to an appropriate newsgroup explaining our concerns and asking for a reason why Twain ended the book as he had. We received several responses. One was from a Twain scholar in the South. He explained that Twain had invested a large sum of money in a new technology. The venture failed, and he was left bankrupt. At the height of his popularity, he was able to contract to write a serial in a weekly magazine. Twain apparently did not have his heart in his work and was somewhat disgusted by what he was doing, so he ended the tale mid-story. My students were surprised to find that in life practical matters often supersede artistic considerations. We next received e-mail from Mark Twain. When quizzed about his authenticity by the students, Twain explained that he had gone out with Halley's comet in 1910 and returned with Captain Stormfield when the comet returned in the mid 1980s. Behind the scenes, I received e-mail from William Gathergood, an Ohio educator, who said he enjoyed "becoming" Twain online for students. For the next several weeks, my entire class became enthralled with Mark Twain as letters were exchanged at a furious pace. Even the students who had not originally read the book were hooked on Twain. Bill told me that he also liked to become Shakespeare online, and to this day, he is captivating students of all ages with letters from Shakespeare that bring the bard's work to life in a way previously unknown. After corresponding with Shakespeare online, I began to notice the appearance of new epithets in my classroom. When annoyed by a peer, one child disdainfully exclaimed, "You loathsome toad!" It is obvious that the words of Shakespeare have real meaning for her.

Another time, one of my students became interested in cockney rhyming slang. As often happens, this child's inquisitiveness made the rest of us want to know more. Children are naturally curious, and I believe it is incumbent on teachers to capitalize on this innate characteristic. The interests of a child can easily be worked into the required curriculum. The language arts curriculum is based on teaching children to read, write, listen, and speak effectively. When children are supplied with a rationale for wanting to learn those communication skills, their enthusiasm becomes a springboard for all sorts of integrated learning experiences. As with Twain and Shakespeare, we turned to a newsgroup for assistance in gathering information about cockney rhyming slang, a coded language invented in the 19th century by cockneys in London whose goal was to speak in front of the police without being understood. In cockney rhyming slang, selected words are coded into two-word phrases that rhyme with the original word. However, only the first word of the phrase is commonly used in conversation. Thus, "bread and honey," which rhymes with money, becomes simply "bread," a slang term for money that is in common use today. Other cockney expressions include "dog and bone" for telephone and "apples and pears" for stairs. One might fall down the apples in an attempt to answer the dog in a hurry.

We began to collect rhyming slang from many sources. E-mail told us warm stories about memories of grandfathers using certain phrases, which we added to the dictionary we were creating. Many requests came in for my students to share their findings. Their dictionary was uploaded to those who wanted it and was posted to a newsgroup, providing a chance for my students to create a product for a real-world audience. We even received a letter by regular post ("snail mail") from Australia giving us examples of Aussie rhyming slang, which differs slightly from cockney slang. What I liked best about this project was that I was learning right along with my students. A teacher's or student's enthusiasm is infectious and serves as a model for lifelong learning.

Consider the teacher in Nebraska who has seven students in her entire multiage school, which is geographically isolated. Her e-mail connection to the outside world allows her students to have contact with other children in faraway places. Her students learn that their very small sphere is part of a much larger one. As my students participated in e-mail exchanges with hers, new understandings about the world around them emerged. Both classes benefited from the contact as each learned what the other's realm was like.

Few children learn keyboarding skills until middle school. Because my students were inefficient at entering data, I generally did all the typing for them. At the end of most days, I would have a folder full of children's writings and questions that I had keyed in for them. I found that this endeavor took little of my time in the big scheme of things. No teacher I know of has evenings free from paperwork; that is the norm. Finding time to enter data for my students was not difficult. It became a daily part of my afterschool work. I warned my students that I would enter data exactly as it had been written. Because of this, and because my students were no longer writing for me but for a real purpose, I found that correct capitalization, punctuation, and spelling became important to them. It was one thing for them to turn in sloppy work to me. It was an entirely different story when they were posting messages on the Internet or responding to e-mail. If we want students to care about what they are doing, I believe we should provide them with purposes for caring.

The fact that English is the predominant language of the Internet makes it convenient for us Americans, most of whom speak no other language. It also becomes obvious to students that most people on the Internet are not Americans. English as a second language often means that the subtle nuances of words in context can be misconstrued by those struggling for the appropriate vocabulary word. I chuckle every time I reread a piece of e-mail we received from a Japanese student who proudly told us he was "the best glutton" in his class. Opportunities like that provide an educator with the teachable moment. It helps make the point that word choice is important for a writer to consider, and it makes students more conscious of careful word choice in their own writings. Students, too, become aware of the complexities of grammar when foreign writers use clumsy constructions. We were always mindful, however, of how difficult communicating in a language other than one's own can be and were careful never to denigrate someone's efforts.

Not everyone is eager to assist children on the Internet, though. While the Internet provides new media that is changing how people interact, not everyone sees this as positive. Whenever I initiated contact with a foreign e-mail address on my students' behalf, I would carefully outline, in an introductory letter, our purpose for the contact, the newsgroup where the address had been found, and a short explanation that we were striving to become globally aware and to learn about other cultures via telecommunications. About 85% of the people we have contacted responded positively and enthusiastically, but another 10–12% ignored our requests. A small fraction, though, would write back in a rather strident fashion stating that they did not appreciate receiving unsolicited junk e-mail and that, no, they would not assist us. A return e-mail letter was sent thanking the person for responding. (Thank you letters are another must. It is just plain good manners to thank someone who assists you, even if it is electronically. Sometimes these thank-you letters result in a rather lengthy correspondence as the volunteer continues to write back. In some cases, I have been corresponding with folks for four or more years as a result of a project my students and I worked on together.)

Sometimes a telecommunication project produces an "Aha!" from students. One cold January, my students were lamenting the lack of daylight hours, which led to a discussion of other places where the sun sets far earlier than here in Virginia. We decided to investigate further and wrote to 30 locations around the world requesting the exact time of sunrise and sunset for a specific date. Then my students had to figure out a way to make sense of the data that came in. They quickly decided to create a table showing each location, the times of sunrise and sunset, and the number of hours of daylight. From this, they created a bar graph ordering the data from least sunlight to most. Last of all, they plotted each site on a world map. As everyone was busy working in small groups, one child looked up and said, "Hey! These locations are in order from the North Pole to the South Pole." The room was abuzz with this new discovery. As the students were finishing plotting the data, another student looked up and said, "Wait a minute. If Antarctica has the most sunlight on this date, why is it the coldest place on Earth?" An ad hoc science lesson ensued with a discussion about the slant of the sun's rays, the tilt of the Earth, and the seasons. As so often happens, a student's question provided an opportunity for the entire class to learn a science concept.

In 1993, each school in Montgomery County was given a BEV account and a more sophisticated computer, housed in the school library. These acquisitions provided new opportunities. The WWW was now accessible, further expanding available electronic resources. About the same time, CD-ROMs and laser discs became more prevalent in libraries, and students readily incorporated the use of them in their research arsenal. The next additions were scanners, which permitted students to include real-world graphics into their work, followed by video tele-conferencing and presentation software. The use of these tools became, for many students, a part of their daily lives.

These examples of technology projects give a general flavor for the mesh between technology and constructivism. Each aspect of information technology resources offers its own unique prospects for integration into a constructivist classroom. In the following section, I discuss ways in which specific technologies can be fruitfully utilized by creative teachers. In each case, guidelines and common pitfalls are described, along with several examples of successful projects that have been made possible by access to the Internet through the BEV.

Applying e-mail

The lowest rung on the Internet ladder is e-mail. A minimum that is required to conduct successful e-mail projects is an e-mail account, a phone line, a low-speed modem, and appropriate telecommunications software. Even an Apple IIe computer will function adequately, although newer computers with high-speed modems or T1 or ISDN access are faster. Sharing a phone line can create scheduling requirements so that an attempt to connect will not cause someone using the line to disconnect, thus losing work in progress. Fortunately, the BEV supports offline email, so that a user can download in one batch all email waiting to be read. Conversely, the user can reply to e-mail without being connected and then upload all pending messages in one session at a later time. This places less strain on a shared phone line.

One decision a teacher must make is whether to allow students to send e-mail directly. Because a teacher is ultimately responsible for all outgoing e-mail, caution must be used. There are many ways to deal with this issue. Generally, by middle school, students can have their own email accounts. In Montgomery County, acceptable-use forms, which both students and parents must sign, are required. The ultimate consequence of misuse is loss of e-mail access in the school setting. The question of students keying in e-mail, however, may come down to one of pragmatics. In a class with one computer, it is inefficient to allow students to type in their own messages, particularly when students are not proficient typists. It is important that the teacher type in messages exactly as the students write them. That forces students to consider their audience, the conventions of language, and mechanics. The teacher should retain the right, however, to ask students to revise inappropriate or shoddy work. I always emphasize to students that American schools are harshly judged when American students demonstrate poor communication skills. This also provides the teacher with the opportunity to teach, in a relevant context, language and mechanics skills.

When e-mail is received, a decision must be made about how to preserve it. It is inefficient to store e-mail in a mailbox. Instead, I recommend downloading the e-mail to a disk or printing hard copies, which are kept in a folder. The offline mail reading software commonly used on the BEV makes it easy to save individual messages to disk. So as not to waste resources, printing e-mail should be considered if it will be beneficial, over time, to have a hard copy available for student use. For example, hard copies of our e-mail from Will Shakespeare were included in a unit developed for use by other teachers.

Another aspect of e-mail that is helpful to teachers is the electronic mailing list. Each mailing list has a defined topic, and people interested in that topic may subscribe to it electronically. Anything a subscriber e-mails to the list address is redistributed via e-mail to all the other subscribers. The administration of a list (subscribing and unsubscribing) may be automatic, as is the case with LISTSERV lists, or manual, where a real person manually subscribes and unsubscribes users. Popular mailing lists are often "gatewayed" to Usenet, meaning that everything posted to the mailing list also appears on a corresponding Usenet newsgroup (and vice versa in some cases). One of the most useful sources of information for educators is the KIDSPHERE mailing list, which is chock full of ideas for projects and advice. A list of known mailing lists is often posted to news.answers.

Accruing a roster of reliable contacts can be rewarding, albeit tricky. My method is to scan the Usenet groups that contain the prefix "sci" because one can find posters from all over the world who are connected with universities and who value education. In my experience, such individuals are often eager to assist American students. I avoid the hierarchies "alt" and "talk" because they are informal and more likely to contain contentious material (they generate more heat than light). Most newsreaders allow you to set up "kill files," which automatically screen out any posting from designated "noisy" users. For example, if you knew that user@some.site continually posted contentious or useless ranting material, you could add user@some.site to either your global kill file (to screen out postings in all newsgroups) or, in the case of some newsreaders, to the kill file for a given newsgroup.

The first time I contact a person, I am careful to write a letter that introduces my students and me and gives a careful outline of our project. I delineate exactly how much help we will need and am sure to say where I found the e-mail address. Often I include a statement about my students needing to become globally aware to be prepared for tomorrow's world. Figure 9.1 is a letter I sent when I conducted my first e-mail project, a comparison of geography.

I believe it is important to respond within 48 hours to all received e-mail. Good manners are a must. Prompt letters of thanks written by the students should be e-mailed as soon as possible. If further questions of clarification need to be asked, include them at the end of the thank-you letter. When sending the same e-mail message to a large number of people, much time can be saved by creating a group alias so the e-mail needs to be sent only once. However, when students are seeking assistance from individuals, it is wise to send individual letters, rather than electronic form letters.

A few words of caution are in order. Follow the rules of "Netiquette." An amusing guide is regularly posted to the Usenet newsgroup news.answers with the subject heading "Emily Postnews Answers Your Questions on Netiquette." One example of good Netiquette is not to write in all uppercase letters, which is the electronic equivalent of shouting. Be clear in what you say. Avoid quoting excessive material from the e-mail you receive when you respond to it. If a respondent asks for personal information about your students, do not send it. When sending e-mail written by students, do not include last names, addresses, or

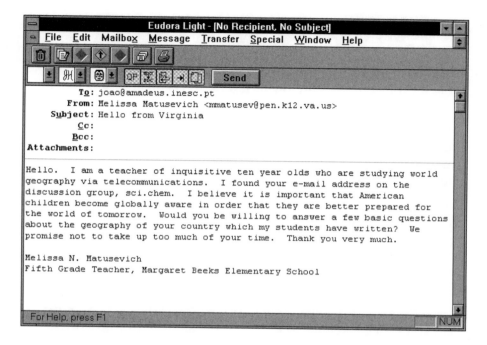

Figure 9.1 A sample letter of introduction for an e-mail geography project.

telephone numbers. Unfortunately, there are all kinds of people on the Internet, some of whom do not have the best interests of children at heart. This is another reason for the teacher to send and receive all e-mail for young children.

High quality e-mail projects provide students with the following opportunities:

- To practice language skills in a real-world environment;

- To make decisions about how to utilize the information received;

- To investigate real-world problems and issues;

- To receive information from many targeted sources about a particular topic;

- To evaluate the quality and relevance of the information received.

Examples of e-mail projects

Teachers are limited only by their imaginations or the questions of their students. E-mail can provide many, many worthwhile learning opportunities for students, as the following few examples demonstrate. I strongly advocate the use of e-mail as an integral part of any classroom program.

Fifth graders at Margaret Beeks Elementary School hosted a national literature contest. Each week a team of students wrote a set of clues about a fictional character. Monday through Thursday, clues were posted as to the character's identity. Monday's clues were hardest; by Thursday, the clues were fairly easy. Students from 31 locations throughout the United States participated, sending their guesses in by noon on Friday. Every Friday afternoon, a list of winners was e-mailed to the participants. The following year, a request came in for a repeat of the contest. Another school volunteered to run it and has done so successfully ever since.

On many occasions I found contacts throughout the world who agreed to assist my students in learning geography. Fifth graders at Margaret Beeks generated a generic list of questions to send to each location. One response provided us with fascinating information when one student suggested the question, "Is there anything particularly interesting or unusual about your country that our other questions have not addressed?" When e-mail from Singapore arrived, the contact told students they might find it interesting that in Singapore it is illegal to have chewing gum, and you can be fined for spitting or not flushing the toilet. At first the students thought this was funny, until we searched for a deeper meaning. We looked at the size of Singapore, the governmental structure, and the population density. Students began to get a much better picture of what life is like there. They came to these conclusions: Singapore is tiny, clean, crowded, and relatively free of crime. The government plays an active role in the daily lives of its inhabitants. The next year, Michael Faye was in the news. Faye, an American teenager living in Singapore, was sentenced to a caning for spray painting graffiti. One of my former students came by and said, "You know, Ms. M., everyone in my class is shocked that Singapore has such severe punishments. But I remembered what we learned about Singapore and wasn't surprised at all."

A fourth grade classroom at Margaret Beeks Elementary School conducted a comparative study of the amount of consumer goods they owned and the goods owned by people in other countries. The families of the

26 students, who are residents of Blacksburg, answered the student-designed survey, telling how many bicycles, cars, microwave ovens, and so on, they owned. Using e-mail, students administered the same survey to 26 people in foreign locations. Taking into consideration that they were communicating with people who had computers and Internet access, students were still able to discern that Americans own many, many times the consumer goods of people in other countries.

Fourth graders at Margaret Beeks Elementary compared the amount of time it takes a letter sent from England by standard post—known as "snail mail" because it is so much slower than e-mail—to arrive in Blacksburg, Virginia, with the length of time it takes a letter to arrive by e-mail. Our English contact mailed us a letter via air mail and then an e-mail message the same afternoon. Students made predictions as to when the letter would arrive and made a class graph of the results. The predictions ranged from three days to two weeks. Already aware of the speed of e-mail, students also made predictions for the e-mail letter and graphed those results. Eight days later the snail mail letter arrived. The e-mail letter required a mere 38 seconds to arrive. One student who was interested in mathematics figured out how many times faster the e-mail was than the snail mail.

The excitement of the Iditarod, the annual dog sled race in Alaska, was brought into fourth and fifth grade classrooms at Margaret Beeks via e-mail. (This was prior to our having access to the WWW through BEV. If I were conducting the project today, I would search the Web first for a more visual impact.) Several times each day, a school in Ohio e-mailed updates of the race. Previously, students had studied the Iditarod and had read the book *Race Against Death*. Students began to root for certain mushers, and the girls, in particular, loved that women sometimes win this test of endurance and commitment. Straight pins with small paper mushers were placed on a map of Alaska containing the Iditarod route. When the updates arrived, students moved the mushers along the trail. The project became more exciting as racers dropped out and the lead changed. The bulletin board provided a visual representation of the information we were receiving by e-mail and added excitement to the project.

At Bethel Elementary School, fourth and fifth graders were studying the American Revolutionary War and the Civil War, respectively. To learn where those historical events fit into the larger scheme of things, students sent e-mail to foreign contacts asking these questions: "What do you believe is the most historically significant world event of the time periods 1760–1790 and 1850–1870? In the time periods given, what is the most historically significant event of your country?" Students made a large wall chart comparing information and were able to draw many conclusions from the chart.

Interactive talk

Interactive talk is a typed conversation between users mediated by client software running on each user's computer. The main difference between interactive talk and e-mail is the immediacy of the former. Interactive talk takes place in real time, unlike the delayed send/response of e-mail. Interactive talk allows dynamic conversations to take place, with the possibility of spontaneous questions. In an interactive talk, a large body of people can pose questions to an expert in a field, much as in a seminar. Interactive talks are more of an interest grabber; in e-mail the respondent may seem distant and abstract, but being able to talk in real time makes the person on the other end more concrete. Students know they are talking to a real person, which reinforces a lot of the conventions of "oral" discourse. When scheduling an interactive talk, students need to be mindful of many of the concepts we strive to teach them. Those concepts become meaningful as students must consider variances in time zones and network connectivity. Losing a connection can reinforce for students that there is a tangible distance between themselves and the other party. A teacher must ensure ahead of time that the talk daemons of each—the software that manages the conversation—are compatible; otherwise, no conversation can take place (rarely a problem nowadays). Figure 9.2 is an example of an interactive talk session.

In the screen capture shown in Figure 9.2, text typed by my class appears in the top half of the window and the replies from our contact are in the bottom. The conversation takes place in real time. Each time someone types dialog on his or her computer, it is appended to the bottom of the previous dialog. Thus, the conversation is much like that on a telephone, but with typing instead of voice. The screen capture illustrates such a conversation after a few minutes of exchange. It is good to use blank lines to cue the other person that you are finished talking so he or

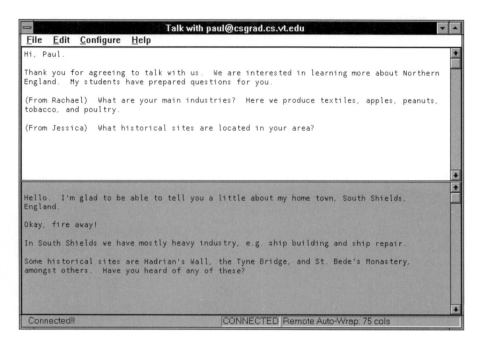

Figure 9.2   An example of an interactive talk session.

she can begin to respond. Some folks prefer to indicate they are finished speaking by typing "(over)," a term used in radio jargon. IRCs are another forum for interactive talk. IRCs must be used judiciously because one never knows to whom one will be speaking. IRCs are online chat groups, so they are less focused. Even though you may join a channel with a topic, the channel topic is not strictly adhered to. The big difference between interactive talk and an IRC is that interactive talk allows the teacher to have a learning goal in mind.

Once, when my fifth-grade students held an interactive talk with a graduate student in Europe, they asked the contact what his schedule was for the rest of the day, after the talk session ended. The student stated that he would go home for his evening meal. My students, who were preparing to go to lunch, understood, finally, what up to then had been abstract notions of time zones and the Earth's rotation. For days afterward, my students would spend time with world maps and globes figuring out time zones in other countries by using lines of longitude and making conjectures as to what people in each country were probably doing at that time.

Usenet newsgroups

Although similar to e-mail in many respects, Usenet newsgroups are different in several important ways that teachers should always bear in mind. The major difference is that in e-mail, the communication is one-to-one, whereas on Usenet it is one-to-many. Instead of a query being posted to a selected individual, it may be presented for the entire world to see. That makes it especially important that the message is not only well written but also sent, that is, posted, to an appropriate place; sailboat enthusiasts are unlikely to be interested in a query about the eating habits of koala bears. When you post to a newsgroup, your message is sent to all sites in the world that choose to receive that newsgroup. Users at each site select to which newsgroups they will subscribe. After a while, your posting will expire; it will not remain on the newsgroup indefinitely. After you post a query, you may receive responses by e-mail, but you should also check your posting for responses for several days afterward as it is propagated to more and more sites.

It is helpful to "lurk," or read, a newsgroup for a while to fully understand its intended purpose before you post any messages. Many newsgroups have "frequently asked questions" (FAQs) files, which not only explain the purpose of the newsgroup but also provide the answers to questions most commonly asked. FAQs appear periodically on the newsgroup, usually once a month. Most FAQs are also cross-posted to the newsgroup, news.answers. One FAQ that everyone should read there is "Emily Postnews Answers your Questions on Netiquette," which outlines the proper way to conduct discourse on Usenet. For any newsgroup, it is wise to familiarize yourself with the contents of the FAQ so as not to post a question that has been answered many times previously.

The WWW

Probably the most exciting portion of the Internet for classroom opportunities is the WWW. Literally thousands of Web pages spring up weekly, providing teachers with a panoply of resources. Not only does the Web provide a massive source of up-to-date information, it also provides a means of publishing one's own information in a permanent fashion. Better still, the type of information that can be published on the Web is far more varied in form than that allowed by e-mail, which is limited largely to text. The Web supports hypermedia—interlinked pages containing text, graphics, sound, and even full-motion digital video—which can easily be viewed by one of the many Web browsers available. With the introduction of Java and the advanced scripting and plug-in features finding their way into the latest Web browsers, the Web is becoming more sophisticated in the possible interactions it allows with users. Instead of being able to merely read a Web page passively, today a user can read and interact with the page in arbitrary ways. For example, it is entirely feasible to design a page to illustrate the laws of motion and mechanics that allows the reader to place objects on an area. The reader could specify the characteristics of each object—mass, density, velocity, acceleration, direction—and then, at the press of a button, simulate the action of all the objects relative to one another.

A helpful tool for information gathering on the Web is the search engine. The Web essentially has no structure. It is a morass of interlinked, individual pages scattered over Internet sites across the world. The Web is now so large in terms of the number of Web pages and gigabytes of information it contains that finding good, relevant information on a topic is not a trivial task. The Web, and the Internet as a whole, embodies the concept of information overload. Fortunately, search engines provide one means of sorting out the electronic wheat from the chaff.

A search engine is the user interface to automatically generated indexes of Web pages. (Automatic indexer is probably a more accurate term, but search engine has caught on in common use.) On a regular basis, a search engine visits all the sites it currently knows about and indexes the Web pages it finds there. If a page visited at a site contains links to other pages or sites, those are added to the search engine's workload, to be indexed in due course. In that way, the search engine mechanically simulates the methodical following of all hyperlinks on Web pages. (Unlike people, the search engine never gets tired or bored.)

The level of indexing performed on each page varies according to the search engine. Some search engines index the complete text of the page, some only the first 100 words or so, and some only the title of the page. All search engines, however, attempt to extract keywords from the Web pages they visit and maintain a global index of all keywords discovered across all Web pages visited. The huge value of all that labor is that the user can then search the index for keywords of interest. Via a simple interface, the user can search by keyword every Web page visited by the search engine. The results of a search are usually in the form of a list of "hits" of the keywords—a list of all those pages indexed that include the search keywords somewhere in them. Each hit listed is usually presented as a hyperlink; clicking on it will take the user directly to the page found by the search (if the page still exists).

Some of the more popular and extensive search engines include:

- Alta Vista (http://altavista.digital.com/);

- Excite NetSearch (http://www.excite.com/);

- Lycos (http://www.lycos.com/);

- Web Crawler (http://www.webcrawler.com/);

- InfoSeek Guide (http://www.infoseek.com/);

- OpenText (http://www.opentext.com/).

All allow a user to search for information on the WWW.

A counterpart to search engines is the manually generated hierarchical topic index. The Yahoo site is a good example. The main difference is that of human intervention, which means that site maintainers decide on a given classification hierarchy, subhierarchy, and so on, and order known sites of interest within it. Many such sites also allow users to submit their own pages to be added to the hierarchy. The main value of this approach is that human beings make the decision as to how to organize available information, allowing some form of quality moderation (i.e., poor sites can be weeded out). A searcher can easily follow down a hierarchy to locate a specific topic.

Search engines do not discriminate as to the age of searchers or the material they index. Because the WWW is an anarchy, it is possible that pages deemed unsuitable for the public school setting will be returned as the result of a search. For that reason, teachers of young children should conduct searches ahead of time and mark the pages to which students should have access. An added advantage is that the teacher can eliminate information overload instead of the students having to wade through marginally useful information. For older students, the previously mentioned acceptable-use contract can limit student access to unsuitable materials, with loss of computer privileges the consequence of unacceptable access. In any large public library, many materials exist that are targeted for an adults-only audience. Just as we would not keep children out of public libraries, neither do we want to deny them access to the rich world of information available on the WWW. If you have a slow Internet connection, you might want to turn off the automatic loading of images, which can often be extremely large and time consuming to download. Also, some Web browsers, such as Lynx, afford text-only access.

I integrate use of the Web into any Internet lesson I design. Recently a colleague, Carol Jortner, and I worked with a mixed group of students from two schools located adjacent to each other. Riner Elementary fifth graders joined sixth graders at Auburn Middle and High School. Our goal was for the students to investigate violence in society and compare what we could learn about violence in the United States with that in other cultures, both Eastern and Western. Students began by calling the court house for information about local statutes on violent crimes and the consequences associated with each. They then searched the Web for further information. The students downloaded a plethora of data, which they had to sift through to decide what information was relevant to their study. They also had to determine the quality of information gleaned and make decisions as to which was most reliable. The next step was to e-mail other countries with specific questions pertaining to violence in their culture. Again, the information had to be assimilated and evaluated. As a result, the students came to many and varied conclusions about the information at hand. A final step was the creation of a slide show using presentation software. Two of the students, Christian, 11, and Whitney, 10, gave a formal presentation at the regional media specialists' conference held at Radford University.

The following example illustrates the usefulness of multimedia. For years, I have been corresponding with a Portuguese student who assisted with a class project many years ago. Joaõ agreed to assist my students in learning about Portugal and even sent us a compact disc of classical Portuguese music. After corresponding with him for a few years, I decided it was time I learned the correct pronunciation of his name. I queried João about how to say his name. For a week we exchanged e-mail with dialogue such as this: "Does it rhyme with 'how?'" "No, it doesn't, but it has a 'zh' sound at the beginning." Finally, in frustration, I wrote, "João, why don't you make a voice clip of you pronouncing your name? Then put it on your Web page." He did just that, and I was finally able to learn how his name should be pronounced. He could also have easily sent me the sound clip via e-mail as a MIME-encoded message.

I have found that perhaps the best reason to use the WWW is that it allows students' work to be presented to a real-world audience. Mark Freeman, a teacher at Blacksburg High School, has posted student work for peer and professional review. Mark's students posted opposing views on set topics such as recycling. People from all around the world had the opportunity to read the essays and agree or disagree using a simple form on the Web page to register their own comments.

I posted student work on a Web page I created for Riner Elementary School. I was careful to include a large group photograph rather than individual pictures of students. I also listed only first names with student work. One third grader, Kevin, had written a story about being hurt in a large department store. His story included details about having to go to the hospital and the store paying the hospital costs. Soon after, e-mail arrived from Denmark commenting on Kevin's mishap and stating how unusual it was to think of anyone paying for hospital care. The sender stated that the government pays for all medical care in his country and that, until recently, it was illegal to charge for medical services. Kevin and his classmates were stunned to learn that someone so far away would read and respond to work posted on their school's Web page. But even more, they were surprised by the different ways cultures provide medical care and assign responsibility for ensuring its access to their citizens.

Again, when students post work on a Web page, their attention to detail, use of language, mechanics, and spelling are improved compared with work they produce for me, the teacher. Like it or not, we teachers do not constitute a real-world audience. Presenting work for strangers makes students naturally more careful about what they write and how they say it. Young children begin to realize that writing is a form of communicating ideas, and if no one can understand what they are saying, they have failed in their endeavor to communicate.

A worthwhile project for a PTA group or parent volunteers is the maintenance of school Web pages. Tools such as Internet Assistant for Microsoft Word and SoftQuad HoTMetaL make the creation of HTML documents as simple as word processing. A special location can be set aside where teachers submit student work for publication. School announcements as well as administrative information, school handbooks, timetables, special projects, and the like can be added as needed. As more and more families acquire access to the WWW, keeping apprised of school happenings will be much easier.

Riner, Virginia, is a small rural community currently in flux. For over 200 years, this farming village has retained its unique identity. Many families have resided in Riner for generations. Now, however, the community is undergoing great change as large subdivisions are being built, and a new mix of residents is moving in. What will happen to the Riner of old? Will it just fade away with no record of its inhabitants and history? A WWW project soon to be undertaken will ensure that does not happen. Principal Keith Rowland, and a cadre of third-grade teachers have agreed to undertake a project that clearly illustrates how active learning can mesh with Internet usage.

The required third-grade curriculum in Montgomery County includes a study of local history. Rather than have students merely parrot the known history they are fed, the third graders will actively add to a body of real-world knowledge through an innovative inquiry-based history project. Students are to discover and chronicle history themselves. As the project progresses, the students will learn many useful skills—interviewing, researching, writing. They also will learn oral presentation skills and the mechanics of creating multimedia to include scanning, creating HTML documents, and digitizing sound and movie clips.

Students will create Web pages to reside on the BEV. They will capture oral histories on tape and then digitize or transcribe them. They will make film clips so older residents can be seen and heard describing life in Riner throughout this century. The students will write the biographies of historical buildings and scan photographs of those buildings. Mathematics will be included as students create scale drawings of floor plans of the old buildings. Students will research archival material at the court house to trace the histories of the buildings. They will create an image-sensitive map of Riner so that clicking on a site will take the reader to a Web page with more detailed information. Instead of passively memorizing facts, students will create the dynamic living history of Riner, Virginia. The project may take several years to complete since it encompasses an ongoing body of knowledge. Successive third-grade classes will be able to build on their predecessors' work. The history of Riner will then be accessed throughout the world. The use of multimedia can enhance the history project and give it greater impact and educational relevance. Descriptions of a few of my educational projects appear on my projects page on the WWW (http://pixel. cs.vt.edu/melissa/projects.html).

Video teleconferencing

Video teleconferencing is the visual equivalent of interactive talk. Instead of sharing words in real time, sound and video are exchanged. Video tele-conferencing opens up possibilities not afforded by interactive talk. Instead of attempting to describe an object, a user can simply show it. Regional accents and state and national anthems can be experienced directly and in real time. American students studying a foreign language can teleconference with foreign students studying English. World events can be discussed from the viewpoints of various cultural groups. Visual and auditory clues missing from e-mail come to life. The low frame rate currently available limits this type of telecommunication, but improvements on the horizon will provide teachers with an exciting means of involving their students in real-time exchanges.

I have used video teleconferencing in a project designed for fifth-grade students at Riner Elementary School to compare the geography of southwestern Virginia with that of the coast of Washington State. We used the CU-SeeMe program to provide an Internet video teleconference link between the two schools. Obstacles that had to be overcome, such as slow connections or the loss of connections, interfered with the quality of the exchange. However, students did learn much about both teleconferencing and geography by participating in the study. Their eagerness and excitement in comparing geographic data were worth the difficulties. Geography became relevant for students as they took a keen interest in talking directly to the inhabitants of the area they were studying instead of merely gathering the information from textbooks. They were able to become actively involved in the study of geography by developing their own questions rather than relying on the limitations of static textbook information.

Another method of teleconferencing is Internet phone telephony, which differs from video teleconferencing in that only sound is conveyed. That limits its usefulness since the visual component of video teleconferencing adds much to an exchange. However, the audio quality in Internet phone telephony is usually of a higher quality because it requires less bandwidth. The higher quality is useful for students conversing in a foreign language, since precise nuances of pronunciation may be important.

Can my school system do this?

By now you are probably thinking, "I can do this," and you can. None of the examples given here are beyond the resources of anyone with a full Internet connection. More and more teachers are becoming aware of the power of the Internet and telecommunications as tools of education. Unfortunately, they often are limited by lack of administrative support and funding. As delineated in this chapter, much can be achieved with modest resources. Getting the administration interested can be relatively easy, as well. Time and again, I have assisted, via e-mail, teachers who are trying to convince their principals that an Internet connection is a worthwhile addition to the classroom. Every day, multiple requests appear on KIDSPHERE, an international newsgroup for teachers, asking for "hello" messages from around the world. Most likely, the requesters print and then display the e-mail replies in a prominent place for the rest of their school to view, demonstrating that the school has access to information spanning the globe. I regularly respond to these messages and often bounce them to other folks around the world who are always eager to assist educators. The next step most teachers take is to design a simple project or to participate in a project posted on KIDSPHERE. When principals understand that real and relevant learning is taking place, they readily support their teachers in the integration of telecommunications into the curriculum. The bottom line is that while this all may seem to be gee-whiz stuff of the 21st century, it is based on simple technology that all schools can afford or already have.

Technology can be a positive force when applied appropriately. The constructivist paradigm is a fitting framework for the effective integration of technology into classrooms and the required curriculum. Over time, systemic change is needed, however, because most teachers are skilled at using a didactic approach, which generally is not compatible with effective integration of technology. As teachers move away from the lecture method to viewing students as active learners and providing for their learning needs in new ways, technology will become a natural and integral part of the classroom, just as it is in everyday life. Many teachers are already making the shift to this dynamic mode of educating students. More and more of their classroom day is spent having students use technology and the Internet as effective means of gathering, processing, and presenting data. As my friend and colleague, Carol Jortner, said to me, "You know, Melissa, I now use the Internet as a part of every single lesson I teach."

References

[1] Collins, A., "The Role of Computer Technology in Restructuring Schools," *Phi Delta Kappan*, 73, September 1991, pp. 28–36.

[2] Dwyer, D., C. Ringstaff, and J. H. Sandholtz, "Changes in Teacher's Beliefs and Practices in Technology-Rich Classrooms," *Educational Leadership*, 48, May 1991, pp. 45–52.

[3] Jonasson, D. H., "Evaluating Constructivist Learning," *Educational Technology*, 31, 1991, pp. 28–33.

[4] Senge, P., *"Applying Principles of the Learning Organization,"* Part II (teleconference), 1995. (Available from Innovation Associates, Inc., 3 Speen Street, Suite 140, Framingham, MA 01701.)

[5] Hodas, S., "Technology Refusal and the Organizational Culture of Schools," *Educational Policy Analysis Archives*, 1 (September 1993). (Online serial available on the World Wide Web: http://info.asu.edu/asu-cwis/epaa/hodas.html.)

[6] Barr, D., "A Solution in Search of a Problem: The Role of Technology in Educational Reform," *Journal for the Education of the Gifted*, 14, 1990, pp. 79–95.

[7] Strommen, E. F., and B. Lincoln, "Constructivism, Technology, and the Future of Classroom Learning," *Education and Urban Society*, 24, August 1992, pp. 466–476.

10

Community Network Technology

by Luke Ward

THE BLACKSBURG ELECTRONIC VILLAGE stands as a model of what can be done with technology to develop a community network, an information utility that can aid education, spur economic development, and enhance a sense of community by increasing the exchange of information among its members. What is most impressive about the BEV is the level of local activity on the network, the degree to which the area's residents have taken this new medium to heart. The technological infrastructure that has been assembled to support the BEV is truly formidable as well: hundreds of modem pool lines and hundreds of direct high-speed Ethernet connections in schools, libraries, apartment complexes, and businesses.

Is this something that can be done elsewhere? Is the new network technology still so esoteric, so expensive that to undertake such an effort requires the direct assistance of a major research university, such as the BEV had from the beginning with Virginia Tech?

About an hour south of Blacksburg, a few minutes north of the Blue Ridge Parkway, a new business has appeared on the Internet. Citizens InterNET Service (http://www.swva.net/) runs a modem pool based in Floyd County, Virginia [1], and currently supports access with local numbers in nine different southwest Virginia counties. All of bucolic Floyd County has exactly one stoplight, which can be found strategically mounted above the junction of the county's two main highways in the heart of beautiful downtown Floyd, hard by the venerable county courthouse, just about a half block from the Floyd Country Store, where a mountain music jam session takes place every Friday night. Floyd residents and businesses enjoy just as much access to the information superhighway as folks in much more urban settings, accompanied by a very different quality of life.

A few hours farther south lies the small town of Abingdon, Virginia (pop. 7,027). Abingdon was first settled back in the mid 1700s, around the time Daniel Boone passed through, and is thick with about as much tradition and history as can be found anywhere in the southern Appalachian region. But there is a strange new addition in town: A funny looking cable leads out of the town hall, dips into the Washington County Public Library about 1,500 ft down the street, emerges again and stretches another 1,500 ft down to the Johnston Memorial Hospital. It is Abingdon's new fiber optic backbone, delivering high-speed Internet connectivity to the heart of town [2]. But look closer—are those plastic cable ties holding that thing up? This cable was only the beginning: By 1998 this network expanded into a full-scale town data service that provides direct Ethernet connections to 50 or more paying business and residential customers throughout the core of old Abingdon.

If Abingdon can do it, any town can. The technology that required the participation of partners like Virginia Tech in 1992 has matured and become much more accessible. If a local government can afford to invest in a new backhoe or dump truck, it can afford to invest in an Internet access infrastructure that can aid economic development, education, and community quality of life.

This chapter is an overview of the Internet networking technology that might be applied to the task of establishing a community network; it is not intended to be a technician's bible with a complete and precise presentation of every issue and task involved. The objective is to give planners, funders, and entrepreneurs the basic background they need to find local partners, work with vendors, understand the technicians, and make rational decisions about what technology options make the most sense for a particular community.

We begin with a short orientation tour of what Internet technology is and how it basically works. Following that is a concentrated, optional dose of computer networking fundamentals. You may want to scan the headings in this section, then refer to it later as necessary to decipher a particular acronym or type of network technology mentioned elsewhere.

After that, we turn to how you can bring Internet access to your town and then distribute that access in the local area. Then, with the communications network infrastructure in place, we survey what network server applications your community might want to run or be required to run.

Finally, we describe a bit about the evolution of the BEV, what is likely to happen with Internet technology in Blacksburg in the near future, and what might be done to take advantage of fiber optic technology to spread direct Ethernet connectivity throughout a small town.

Basic Internet technology

What is the Internet?

The Internet is the aggregate interconnection of many computers on many computer networks all over the world, whose users exchange information using a common set of communication tools.

About 15 years ago (a century in computer years), it became common to connect a number of computers in a single building or a small area for the purpose of sharing information among them. Such a configuration is known as a local area network (LAN). As time went on, it became desirable to link LANs across longer distances, forming what is known as a wide area network (WAN). The Internet is the biggest WAN there is, comprising about 80,000 networks by 1996 [3]. All the computers connected to those networks, by 1999 at least 40 million, possibly more than 60 million, can communicate with each other [4].

Any computer, whatever type it is, that is connected to the Internet can communicate with any other connected computer. It does not matter if it is a PentiumPro running Windows95, a brand-new 400-MHz G3 PowerMac, or a venerable water-cooled IBM mainframe running OS/390—they all can communicate with each other over the Internet because they use the same set of official Internet protocols, computer communication languages and procedures. That is the basis for the openness of the Internet environment. The Internet protocols are officially and precisely described in publicly available documents called Internet standards [5]; using those standards, any vendor can develop software or hardware that implements the Internet protocol suite for any operating system running on any hardware platform.[1]

Laying aside the well-worn analogy of an information superhighway for a moment, consider the similarities between the Internet and the telephone system you use every day. The phone in your home, along with a number of other phones in your community, are all connected to a telephone company switch in a building somewhere reasonably nearby, something like a computer LAN. To call another telephone, you pick up your receiver, wait to hear the dial tone, key the number of the phone you want to call, listen to it ring, and wait for the person at the other end to pick up the receiver. This standard sequence is something like that defined by the protocols computers use to exchange information over the Internet: One computer starts the connection process, specifies the address (like an Internet phone number) of the computer to be contacted using a standard language of electrical impulses, waits for the remote computer to respond, and so forth.

Further consider the fact that you can call a telephone that is hooked up to an entirely different phone company on the other side of the country. That is something like the Internet WAN. Your phone company has an exchange point where its lines are connected to those of other phone companies. You can even call a phone in a different country or one manufactured by a different company. All of this happens relatively transparently—all you do is dial a number.

---

1. There are many standard Internet communication protocols. The most important of these, the one that defines the fundamental addressing and routing structure, is simply known as "the Internet Protocol" or "IP."

How does the Internet really work?

To get a better idea of exactly how Internet addresses work and how the networks are connected, let's follow what happens when a computer on one side of the continent requests information from a computer on the other side (Figure 10.1).

John in Grays Harbor, Washington, hears about the BEV from a friend. "Take a look at it," she says. "Their home page URL is http://www.bev.net." John goes home, fires up his trusty PowerMac and modem, dials into the modem pool of the local Grays Harbor Internet service provider (ISP) to start an Internet connection, and carefully types the uniform resource locator (URL), a World Wide Web (WWW) address for an individual page or service, into his Web browser. (Examples of Web browsers are Mosaic, Netscape Navigator, Microsoft Explorer, and CyberDog.) The browser and other IP programs on the PowerMac form his request for the BEV home page into one or more IP packets, strings of data tagged with the address of the sender (John's PowerMac) and the destination (the BEV Web server).

The packets are sent through John's modem over the phone line to the Grays Harbor ISP's modem pool. From the modem pool, the packets pass over another phone line, a special high-speed data connection, on to the

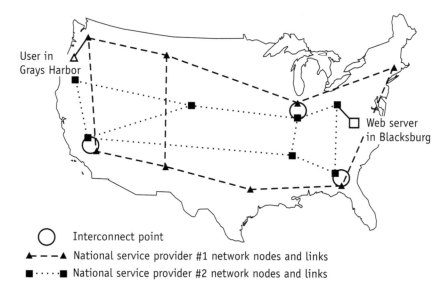

Figure 10.1 Internet communication through multiple NSPs.

network of one of the national service providers (NSPs). NSPs function something like the long-distance phone companies in that they provide network service reaching across the country and the world. (In fact, several of the NSPs are long-distance phone companies.) In this case, the Grays Harbor ISP's link to the Internet is its connection with an NSP.

As it turns out, the Grays Harbor modem pool ISP is connected to a different NSP than the one to which the BEV is connected. That complication is negotiated by passing the packets to an interconnect point, also known as an exchange, peering, or network access point (NAP). This is an ultra-high-volume junction where multiple NSP and regional ISP networks meet to exchange packets. John's packets from Grays Harbor reach the interconnect point and pass onto a new NSP network, then on down to the BEV network and the BEV Web server.

The BEV Web server computer runs programs that implement the Internet protocols, which interpret the arriving packets. In response, out onto the BEV network go packets containing the text and images that make up the BEV home page, each IP packet now labeled with the Internet address of John's PowerMac as the destination and with the BEV Web server as the sender.

The original address is the "www.bev.net" part in the URL that John's friend told him about. The trailing "bev.net" identifies a network domain, and the "www" identifies the main Web server for that domain. This particular type of Internet address is known as the fully qualified domain name of the computer running the BEV Web server. It corresponds to a second address, a unique numeric code, 198.82.165.36, which is known as an IP address. This is the address that is encoded on the IP packets passed through the network. All along the path an IP packet travels, devices called routers form junctions between alternate communication links and pass the packet on according to its destination IP address.

The key that makes it possible to route the trillions of packets among the millions of computers attached to the Internet is IP, which defines the packet structure and addresses. This is the foundation on which the entire fantastic edifice is built—to a large extent, IP *is* the Internet.

The problem of being too successful

When all the machinery is working fairly well, the sequence of query and response can happen in a few seconds. Unfortunately, that is not always the case—the incredible growth rate of the Internet is pressing the system's current limits.

Many parts of the Internet are redundantly implemented to ensure that the packets still get through even if there is a failure. NSPs and ISPs routinely provide alternate connection paths in their networks, so routers can send packets by another link when the normal one fails or is congested with too much traffic. The domain name server (DNS) system that translates domain names into IP addresses is so critical that it is implemented in a completely redundant and distributed fashion, so if one server fails or is unreachable another can be used. The Internet communication protocols themselves include automatic retries, so if a packet vanishes along the way, another is automatically sent to replace it.

All these safeguards are currently being stretched, sometimes past their breaking point: Traffic on the Internet has been increasing so quickly that it has been difficult for the ISPs, NSPs, and interconnection points to keep up. By mid-1999 one of the busiest interconnects, MAE-East, located in the Washington, D.C., area, was handling data from over 100 different connections for almost as many different telecommunication vendors and organizations at an aggregate peak of about 2,200 Mbps [6].

Increased speed in transmission rates translates into increased packet capacity. Packets arriving simultaneously can be buffered or held by a router for a second or two while a backlog clears. However, if all possible paths are too congested, buffer overflow occurs and the packet is dropped. When that happens, the computer originating the packet will resend it, but not indefinitely; as congestion increases, users will see increasing delays and eventually may receive a "server not responding" message.

Access providers are competing on the basis of reliability and are increasing capacity as fast as they can. More interconnect points are also coming online, but with the current rate of growth the dropped-packet problem will continue to be a nuisance for the foreseeable future. In this climate, it pays to choose a provider carefully.

Another problem is the growing shortage of IP addresses. Part of an individual computer's IP address identifies the Internet network to which it is attached, and the number of Internet networks is increasing rapidly. This not only threatens to exhaust the supply of available addresses but also compounds the complex problem of tracking routes for all the networks.

New tools like dynamic IP address allocation and network address translation (NAT) have stretched the existing supply of addresses, but a new address system for IP will be required eventually [3]. The next-generation IP (IPng, also known as IP version 6, or IPv6) promises by the most conservative of estimates to provide more than 1,500 IP addresses for every square meter of the planet, and a number of trial implementations are now being developed. A key objective is compatibility between the old and new versions of IP; the hardware and software you buy today should continue to be usable as IPv6 is deployed.

A concentrated dose of LAN, WAN, and TCP/IP fundamentals

To complete our introduction to the Internet and how it is delivered to your door, this section surveys the technical underpinnings of network technology in more detail. If you are not inclined to delve that deep at this point, skip this section and go on to "Bringing the Internet to your town" (page 237). The headings here are intended to make it easy to explore specific items on an as-needed basis.

Bits and bytes, Kbps and Mbps

The data communicated between computers can be text, images, sounds, or video, but they are all encoded in digital form, that is, sequences of bits. A bit is a single 1 or a 0, and a byte is a sequence of 8 bits. Each 8-bit byte is generally used to encode a single printed character, but because of the overhead required in data communications, 10 bits per character is a convenient rule of thumb to use when talking about network transmission rates. Data rates, a measure of capacity or bandwidth, are typically expressed in terms of thousands of bits per second (Kbps) or millions of bits per second (Mbps).

In telecommunications, the K in Kbps really means 1,000, and 1 Mbps is 1,000 Kbps. In computer science, however, when applied to disk space or random access memory K means 2 to the 10th power, or 1,024, and 1 megabyte is 1,024 kilobytes.

Compression is often used to encode data in a more compact form to save transmission time and to deliver it at an effectively higher speed. Compression works particularly well for text: Modern techniques can often compress text data to as little as 25% of its original size. That is how modems that connect at 28.8 Kbps over the phone line can potentially deliver data at 115.2 Kbps to the computers that use them. Images, however, are generally already stored in compressed form, and it is not possible to compress them much more for transmission. The Web pages typically downloaded over Internet connections are unfortunately mostly made up of images.

How many bytes are a lot? The uncompressed full text of *Moby Dick* comprises about 1.2 million bytes, roughly the capacity of a 3 1/2-inch floppy disk. A typical image occupying about one-half of a Web page is about 75,000 bytes. A CD-ROM disk can hold about 650 Megabytes.

Internet domain names and addresses

Just as every telephone has its own number, every computer attached to the Internet has its own address. This address, however, comes in several forms.

The most familiar type of Internet address is the fully qualified domain name that appears in the World Wide Web addresses now seen almost everywhere, for example, the www.bev.net in the http://www.bev.net/ URL of the main BEV Web server. The "www" part identifies a particular computer, or host, and the "bev.net" part identifies a network domain, a group of addresses associated with a particular business, organization, or other institution. Domains can be further subdivided with additional qualifiers; for example, the Virginia Tech "vt.edu" domain is further subdivided by departments, so an individual computer in the computer science department might have a domain name like alpha.cs.vt.edu.

The right-most qualifier in the domain name ("net" in the case of "bev.net") is the root domain. Different root domains are used for different categories of groups: "net" is used for network service providers, "edu" for four-year colleges and universities, "com" for businesses, and so forth. Localities and school districts in the United States are now associated with the "us" domain; examples are blacksburg.va.us for Blacksburg, Virginia, and montgomery.k12.va.us for schools in Montgomery County, Virginia.

IP addresses, networks, and routers

The bev.net domain happens to be used to name only computers in Blacksburg, but some domains, such as those of large corporations, span the globe. For instance, given the domain name www.ford.com, it is not immediately obvious where in the world that particular Ford Motor Company computer might be.

That problem is managed with a second type of address, the IP address, a unique numeric code that is actually used to specify the sender and the destination on each IP packet. Every fully qualified domain name corresponds to a numeric IP address; www.bev.net, for instance, corresponds to 198.82.165.36. Translating the domain name into the IP address specifies where to send a packet destined for that machine: The leading part of the IP address (198.82.165 in the example given here) identifies a specific physical network at a specific physical location.

As previously mentioned, routers stand at the junctions between different Internet communication paths. When an IP packet arrives on one path, the router must decide onto which path to pass it next. The router does that by examining the network part of the destination IP address and comparing it against a routing table that defines which paths lead to which physical networks. All the routers work together to keep their routing tables up to date by exchanging routing information using special routing protocols.

Because they operate at the IP level and can closely control which packets are passed, routers are most commonly used to terminate relatively slow and expensive WAN links such as T1 lines. For the same reason, they are also instrumental in protecting internal LANs against unauthorized access from the Internet at large.

The domain name/IP address mechanism is also useful when services are moved from one computer to another. If Ford Motor Company decides to move its Web service from a computer in Detroit to one in Blacksburg (or even just across the hall), all it needs to do is link www.ford.com with a new numeric IP address. Users can continue to use the same familiar alphabetic domain name.

Domain name service

The domain name service (DNS) keeps track of which domain name goes with which IP address. DNS is so critical to the function of the Internet that it is operated in an entirely distributed and redundant fashion. At least two name servers are designated as being the source of authority for a given domain; one acts as the primary name server and defines the correct information, and the others act as the secondary name servers, automatically replicating the primary server's data and acting as backup sources in case something happens to the primary machine.

The domains themselves are managed in a hierarchical fashion. The root name servers maintain information for the root domains (e.g., "net," "com," "us," and "edu"), while authority for the subordinate domains (e.g., "vt.edu," "bev.net," and "blacksburg.va.us") is delegated to other servers. The BEV name server could in turn delegate authority for a research.bev.net domain to another name server.

For a domain to be used, it must be registered in the root name servers. That task is currently managed by the U.S. Domain Registry for the U.S. domain (http://www.isi.edu/us-domain/) and by a variety of providers for other domains (http://www.iana.org/domain-names.html). See those services' sites for policies on domain names and information on how to register.

TCP/IP and protocol stacks

Once the packet has been marked with the IP address of the sender and the receiver and the data have been packed inside, the problem for the sending computer becomes: What will carry the packet? IP itself does not define patterns of electrical impulses or flashes of light; that is the business of lower-level protocols, which define their own boxcars (at that particular level, often called frames) into which IP packets are packed.

The upper-level/lower-level relationship in which one service accomplishes its assigned tasks by employing the services of another is common in computer communications and can conveniently be represented by the slash (/) shorthand. For example, IP/Ethernet specifies IP packets conveyed over an Ethernet network. Things being what they are in the computer realm, a trick that works once is replicated indefinitely, leading to such babble as RFC822/SMTP/TCP/IP/Ethernet. A sequence of these building blocks is referred to as a protocol stack.

In general, each level of protocol adds its own contribution to the overall communications solution. For instance, IP provides basic packet addressing and routing capabilities. Most real Internet services, however, rely on the combination of transmission control protocol running over IP (TCP/IP). TCP adds reliability to the data transfer, correcting errors and resending IP packets as necessary, providing a reliable data stream service to the applications above, which in turn use TCP as a lower-level protocol.

There are a variety of ways to transport IP packets. Different low-level networking technologies each have their own advantages and disadvantages in terms of speed, distance, and other characteristics and hence are chosen depending on the requirements of the situation at hand. We will consider some that are commonly encountered (SLIP, CSLIP, PPP, Ethernet, WAN telecommunication links like T1) and briefly mention a few other options, including ATM, wireless, cable modems, and DSL.

SLIP, CSLIP, and PPP

Serial line Internet protocol (SLIP) and point-to-point protocol (PPP) are two protocols for transmitting IP packets over a serial connection such as that provided by a common modem. They specify special signaling characteristics that are needed for that kind of link. PPP and compressed serial line Internet protocol (CSLIP) also abbreviate the IP packets as much as possible, since modems are relatively slow. Throughput generally will be much improved if CSLIP or PPP is used instead of straight SLIP, especially when used with an application that generates many small IP packets, such as Telnet.

Common modem standards and speeds are V.34 running at 28.8 Kbps and V.90 running up to 56 Kbps. Distance is limited only by the extent of the telephone system, although data rates can fall dramatically over scratchy long-distance connections. V.90 modems in particular push the envelope of what can be done with standard voice telephone lines. For instance, even when connecting to a modem pool just across town, users often complain that V.90 modems seldom seem to deliver more than 48 Kbps.

Common data communication rates are listed in Table 10.1. Note that some services are asymmetric—they deliver a higher speed downstream from the service provider to the user than they do upstream.

Table 10.1
Common Data Communication Rates

Connection Type	Data Rate	Multiple of 28.8 Kbps	Notes
V.34 modem	28.8 Kbps	1.0	Compression can increase rate for text.
V.90 modem	56 Kbps	1.9	Compression can increase rate for text; often limited to 48 Kbps or less by line quality.
DDS*	56 Kbps	1.9	—
ISDN BRI†	128 Kbps	4.4	—
Fractional T1	64 Kbps to 1.472 Mbps	2.2 to 51.1	In multiples of 64 Kbps.
ISDN PRI‡	1.472 Mbps	51.1	23 64 Kbps data channels
Frame relay	64 Kbps to 1.536 Mbps	2.2 to 53.3	Committed rate typically is half the burst rate.
T1 (DS1) leased line	1.536 Mbps	53.3	24 64 Kbps channels
DSL leased line	256 Kbps to 6 Mbps down, 64 Kbps to 2 Mbps up	8.9 to 208	Varies with equipment, distance
10BASE-2 Ethernet	10 Mbps	347	RG58 coax cabling
10BASE-T Ethernet	10 Mbps	347	Unshielded twisted pair (UTP) cabling
10BASE-FL Ethernet	10 Mbps	347	Up to 2,000m with multimode fiber
Cable modem	100 Kbps to 30 Mbps down, 40 Kbps to 1 Mbps up	3.5 to 1040	Varies with supplier configuration, user activity.

Table 10.1 (continued)

Connection Type	Data Rate	Multiple of 28.8 Kbps	Notes
Wireless line-of-sight	1.5 to 30 Mbps	52 to 1040	Speed and distance vary with product
T3 (DS3) leased line	43.008 Mbps	1490	—
100BASE-TX Ethernet	100 Mbps	3470	Cat-5 UTP cabling
100BASE-FX Ethernet	100 Mbps	3470	Up to 2,000m with multimode fiber
OC3 SONET§	155 Mbps	5380	—
ATM‖	25 to 622 Mbps	868 to 21,600	155 Mbps up to 2km with multimode fiber
1000BASE-CX Ethernet	1000 Mbps	34,700	25m with special coax.
1000BASE-T Ethernet	1000 Mbps	34,700	100m Cat-5E or Cat-6 UTP
1000BASE-SX Ethernet	1000 Mbps	34,700	275m with multimode fiber.
1000BASE-LX Ethernet	1000 Mbps	34,700	5km with singlemode fiber.

*Digital dataphone service
†Basic rate interface
‡Primary rate interface
§Synchronous optical network
‖Asynchronous transfer mode

ISDN

Integrated services digital network (ISDN) is a digital telephone service that may be offered by your local phone company. It is newer than plain old telephone service (POTS), but ISDN is old compared to digital subscriber line (DSL) technology. Some proponents herald ISDN as the next-generation replacement for standard telephone modems; others criticize it as "too little too late."

Each user must be located within 18,000 cable feet of their phone exchange where the ISDN-capable switch is located, but ISDN basic rate interface (BRI) service can be used to provide two 64-Kbps data circuits

(128 Kbps total) between local sites or even between locations served by different phone exchanges. Primary rate interface (PRI) ISDN bundles 23 64-Kbps data channels onto a single circuit, and provides service comparable to a T1 leased line. PPP is often used to carry IP packets over ISDN links.

One problem with using BRI ISDN to support large groups of users or heavily used servers is that 128 Kbps is not much more than twice the data rate offered by a 56-Kbps modem. Another problem is the cost: ISDN is a dialed service, and users are often charged a per-minute rate for calls. If you use an ISDN link for long periods, you may find that you might as well pay for a leased line. Check rates in your area; you may be eligible for special ISDN pricing if you already purchase other business services from your phone company.

Because of its dialup nature, ISDN generally is not suitable for sites running Internet servers. It is possible, however, to automatically start an ISDN connection fairly quickly compared to a modem—in some cases only three seconds is required—so if your ISP offers dial-on-demand ISDN service, it may suffice for a very lightly loaded server.

T1 and DDS leased lines

WANs use high-speed data circuits to carry packets between two or more LANs over long distances. It is possible to do this using standard dialup modems, but their relatively slow data rate makes it impractical to serve more than a very few users.

In data telecommunications, a leased line is a specially conditioned telephone company circuit that forms a permanent connection between two sites. There is a setup charge for a leased line as well as a periodic, usually monthly, cost. The fees increase with the data rate of the line and the distance it travels.

T1 leased lines have a circuit speed of 1.544 Mbps, and can carry data at up to 1.536 Mbps (8 Kbps is lost to overhead such as end point synchronization and error detection). A special device called a channel service unit/data service unit (CSU/DSU) transforms digital network signals into a form that has the special signaling characteristics required by T1 lines. The CSU/DSU functions essentially as a T1 modem. Figure 10.2 illustrates the equipment required for T1 leased line connections.

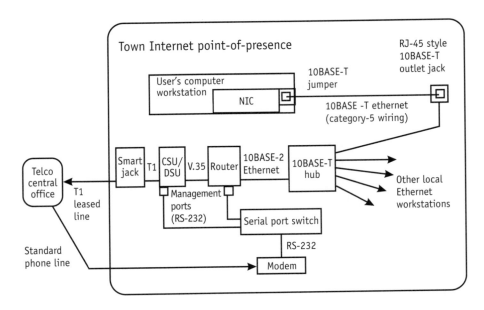

Figure 10.2 Town Internet point-of-presence equipment for T1/fractional T1/frame relay.

The T1 data rate of 1.536 Mbps actually consists of 24 different 64-Kbps data channels. Some CSU/DSUs can be used to split out portions of the data channels available on one T1 line and use them for different purposes. For instance, you could have one T1 line that handles both telephone service and IP data traffic. Using fewer than 24 of a T1 line's channels is called fractional T1. Typical fractional T1 rates are 256, 512, or 768 Kbps.

The leased line that comes into a building actually looks just like a regular two-pair phone line. The phone company typically terminates the line in a box on the wall called a smart jack, which is something like the telephone network interface box on the side of a house where a residential phone line terminates. The smart jack includes some electronics that help the phone company remotely test and monitor the T1 line.

Some length of in-building T1 wiring will run from the smart jack to the CSU/DSU. The CSU/DSU connects to the T1 line on one side and usually has a V.35 synchronous digital connection on the other side. This connection is something like the familiar RS232 serial port found on PCs, but it is capable of a much higher data rate. The V.35 cable plugs into your router.

In addition to the V.35 and telephone line connections, a good-quality CSU/DSU also has an RS232 serial management port. You can plug a PC running a terminal emulator program such as HyperTerminal or Kermit into this port and type commands to reset or configure the CSU/DSU. Routers also often have RS232 management ports.

You can also connect a modem to the router and CSU/DSU serial ports, which will allow you to remotely reset or configure them when the regular network goes down (usually in the middle of the night!) This is accomplished by attaching a modem to a regular incoming phone line, then attaching its serial port to an RS232 remote communications switch; you can find these switches in various telecommunications catalogs. The switch in turn has two ports on the back end, one each for the CSU/DSU and the router. When you call the modem, the switch answers and allows you to select which device—the CSU/DSU or the router—you want to talk to.

Other types of leased lines used in WANs run at other data rates. Digital dataphone service (DDS) is an older type of 56-Kbps connection whose low data rate makes it impractical for many IP applications. Note that a CSU/DSU and router that can handle a 56-Kbps or fractional T1 rate may not be able to handle a full 1.536-Mbps T1. T3 is similar to T1 but carries up to 672 64-Kbps data channels (43.008 Mbps total), and generally costs four to six times as much as T1.

Frame relay and SMDS

Frame relay is a WAN link like T1 or DDS leased lines, but it works in a fundamentally different way. It is not available in all areas, but it can be useful for a link to a distant site since there usually are no mileage charges. Frame relay has a downside, though, in that you share the bandwidth with other customers.

While normal leased lines are real dedicated circuits used exclusively for passing data between two sites, frame relay provides a permanent virtual circuit (PVC) through a frame relay network for each communication link. Each user site has one physical connection to a frame relay network switch, usually one run by the local phone company.

Communicating sites each send frames of data to the switch, which interprets address information on them and passes the frames on to the indicated destination. This is something like the way IP packet switching works, but frame relay uses a simpler, lower level protocol than IP, and virtual circuits are permanently configured between sites.

Frame relay connections are priced according to a committed information rate (CIR), the minimum speed at which the frame relay packet switching service provider promises to transmit and receive packets. Your line can also run up to a burst rate, typically twice the CIR. Thus, you might have a frame relay connection with CIR and burst rates of 512 Kbps and 1.024 Mbps, respectively. Transmission of packets at the lower CIR rate is guaranteed, but packets sent at a higher rate may or may not get through. If the provider's frame relay network is congested with too many packets from too many subscribers, you will seldom see your burst rate.

The frame relay access line is the same kind of data connection used for a leased line. The rate of the line must be sufficient to support the burst rate. Frame relay equipment at the customer end is basically the same as the equipment used to terminate a T1 line (see Figure 10.2.) However, not all CSU/DSUs and routers are capable of running frame relay.

In addition to the lack of access line distance charges, frame relay offers another advantage: A single router port and CSU/DSU connected by a single access circuit to the frame relay network can communicate with more than one other frame relay site. Additional PVCs can be created with configuration changes in the router and the frame relay switch. After your first such connection, that results in an equipment saving of one CSU/DSU and router port for every additional remote site served, although you may need to spend more on a central router with frame relay capability and sufficient power to handle many PVCs.

Although a central router can service many PVCs, each with its own CIR, your access circuit to the phone company's frame relay network cannot run faster than its data rate, usually T1 (1.536 Mbps). Because the network activity on each PVC is intermittent, you can oversubscribe this interface to some extent, provide service to a number of PVCs whose total CIR exceeds the rate of the access circuit. In general, you do not want to oversubscribe the access circuit rate by more than a factor of 2; that is, you do not want to try to service more than six 512-Kbps CIR

PVCs with one 1.536-Mbps T1 access circuit. In such a scenario, if remote sites frequently meet or exceed their CIR, many packets will be lost and users will experience significant delays.

Switched multimegabit data service (SMDS) is another type of packet switch-based network data service that is similar to frame relay, but is generally used at higher data rates (1 to 45 Mbps) over longer distances and is more expensive. SMDS can be an attractive option if you need at least a T1 rate and want to install equipment now that will allow you to gradually step up to higher speeds in the future.

Cable modems

Cable modems are a new technology that supports data transmission over a cable TV network. Some equipment vendors are claiming that speeds of up to 30 Mbps are possible with a delivery range of several miles. A number of industry analysts believe that cable modems will supplant ISDN as the next-generation data delivery tool for residential customers. Large cable system operators have ordered hundreds of thousands of cable modems from hardware suppliers, planning to rent them to consumers.

Much of this technology, however, is still unproven. Cable TV systems were originally designed to carry signals in one direction, from the local cable office (or head end) to consumers' homes. Internet access is a bidirectional activity, and analysts estimate that 50% or more of the cable facilities in the United States cannot yet handle two-way data communications. Even with equipment upgrades, most cable modem systems support only asymmetric data transmission: The user can receive data at a high rate but only send it at a lower speed. Some systems even use the cable only for the downstream data and use a regular modem and separate telephone line for the upstream path.

Other concerns about cable modems stem from the fact that they are a shared-access medium. The downstream channel may have a data capacity of 30 Mbps, but each user on that channel shares that bandwidth. If the cable company loads a single data channel up with hundreds of active users, each user can end up with only 150 Kbps. Also, since every user's transmissions are exposed to every other user's modem, security can be a concern.

Cable modem manufacturers and system operators are addressing these issues. Encryption can be used to guard the privacy of users' communications. Some newer cable modem manufacturers claim to have the ability to reserve bandwidth for each customer, preventing a few high-volume users from monopolizing a channel.

If your local cable service provider can guarantee high-speed bidirectional bandwidth it may have what you need to connect more than just individual users. Check with current customers and run some tests. An upstream rate that never rises above 128 Kbps will not satisfy the demands of an Internet server, and a downstream rate that dwindles to 300 Kbps or less during high-volume periods will not be enough for a site providing Internet access to more than a few users.

Digital subscriber lines

Asymmetric digital subscriber line (ADSL) is one of a whole flock of new xDSL technologies that attempt to move more data faster through plain old copper telephone wires. They accomplish this by using the wire's capability to carry signals outside the range of normal voice bandwidth. With data rates up to 70 Mbps claimed for some future variants, DSL modems constitute the telephone companies' answer to cable modems. As data rates increase, however, DSL distance capability decreases. With 1999 technology you can typically get downstream rates of about 1.5 Mbps up to 18,000 cable feet from the DSL service supplier's equipment, a digital subscriber line access multiplexer (DSLAM).

Like cable modems, several types of DSL are asymmetric, offering a diminished upstream data rate which tends to make them unsuitable for connecting Internet servers. Unlike cable modems, DSL lines are not shared-access—you get all the bandwidth your line can supply, and there is less opportunity for eavesdropping.

Thoroughly investigate your local DSL options before ruling it out as a connection strategy. Because of telecommunications deregulation, there may be one or more competing local exchange carriers (CLECs) offering DSL connections in addition to your traditional phone company, the incumbent local exchange carrier (ILEC).

Ethernet

Ethernet is a fast, inexpensive way to transport IP packets (or any other kind) over short to medium distances, up to about 12 miles. It has been around since Dr. Robert M. Metcalfe invented it at the Xerox Palo Alto Research Center in the 1970s and has since been reinvented a number of times to run at faster speeds, over longer distances, and over different types of cabling (or media) [7]. The current standard defining Ethernet, published by the Institute of Electrical and Electronics Engineers, is known as IEEE 802.3.

Ethernet is fundamentally a shared-access medium; implementations vary, but at root more than one computer communicates using the same cable. Before sending an Ethernet frame, the sender listens to make sure no other computer is already sending. When it happens that more than one computer tries to send an Ethernet frame at the same time, a collision occurs. Both senders detect the collision and interrupt their transmission. Each then waits a random period of time (measured in millionths of a second) before attempting to send again.

This procedure is analogous to the behavior of a polite group at a dinner party. When two persons at the table speak at once, both stop, and in a moment one or the other picks up the conversation again. It all works very well so long as the group size (the number of connected computers) and the table size (the dimensions of the Ethernet network) are not too large. Care must be taken to not overload an Ethernet with traffic and to not violate the limiting rules for the media type in use.

One section of Ethernet cabling is known as a segment, and has a specific length limit depending on the media type. Segments can be connected to achieve longer distances with repeaters that echo signals from one segment to another. One or more connected segments make up a collision domain, a zone in which all senders must be able to detect collisions within a certain period of time. In addition to the segment-length limit, there is a limit on the number of repeaters between any two senders in a collision domain.

Ethernet networks can also be extended beyond collision domain limits with bridges or routers (see the subsection "The differences between repeaters, hubs, bridges, switches, and routers").

Ethernet ARP, NICs, MAUs, AUI connectors, and MAC addresses

Each connection between the network and a computer or a piece of communications equipment (which is more or less a computer at heart) is called an interface. Computers are generally connected to Ethernet with a piece of hardware called a network interface card (NIC). Many new systems now come with an NIC built in. Each and every physically manufactured Ethernet interface has a unique Ethernet address, which is known as a physical, hardware, or media access control (MAC) address. This is not the same as the IP address! A special protocol called address resolution protocol (ARP) is used to associate each NIC's MAC address with one or more IP addresses.

The NIC will have one or more sockets where the actual connection to the network can be made. The most generic type for 10-Mbps Ethernet is a 15-pin attachment unit interface (AUI) connector, which can be plugged into a cable leading to a medium attachment unit (MAU). MAUs (also known as transceivers) in turn provide connectors that vary, depending on the type of Ethernet cabling for which they are designed.

Instead of an external MAU, most NICs now include an internal MAU and simply offer plugs that can be directly attached to one or more different types of Ethernet media. The advantage is a cost savings because the external MAU is not necessary. The disadvantages are the loss of debugging feedback provided by the blinking lights on the transceiver and that a computer moved to a different type of Ethernet may require a new NIC.

The differences between repeaters, hubs, bridges, switches, and routers

A repeater is a device with two interfaces (also called ports, as in "two-port repeater") that can be used to connect Ethernet segments. All Ethernet frames arriving at one segment connection are echoed to the other.

A repeating hub is a multiport repeater, a single box that provides repeater connections for more than two different segments. Traffic arriving at any port is echoed to all the others. Repeating hubs are most often used to aggregate a number of link segments, individual Ethernet connections that run directly from the hub to single MAUs in a star configuration. This is usually the case with 10BASE-T and 100BASE-TX Ethernet.

Since all traffic is echoed to every segment, it is possible for users attached to a repeating hub to eavesdrop on each other's packets with so-called sniffer programs. Some more expensive repeating hubs include security features that mask the data portion of any Ethernet frame not destined for the MAC addresses present on a given segment, but as prices drop Ethernet switches (also known as multiport bridges—see below) are becoming the preferred way to deal with this problem.

Some repeaters are stackable. Normally when one repeater is connected to another, both count against the limit on the number of repeaters between any two MAUs in a collision domain.[2] Stackable repeaters and repeating hubs get around this with some special electronics that allow several of them to be considered as a single repeater when they are located together (usually vertically stacked) and connected directly to each other with cabling provided by the manufacturer.

A bridge is smarter than a repeater; it listens to the Ethernet traffic and, by observing the sender MAC addresses on the Ethernet frames, learns which interfaces lie down the different segments to which the bridge is connected. A bridge will resend Ethernet frames only to segments to which it thinks the interface identified by the receiving MAC address is attached.

An important distinction between bridges and repeaters is that bridges define the boundary of a collision domain, while repeaters do not. Thus, bridges can be used to expand Ethernets beyond the limits imposed for a single collision domain. Repeaters are also limited to connecting Ethernet segments running at the same data rate, while bridges can be used to connect mixed 10-Mbps, 100-Mbps, and 1000-Mbps Ethernet segments.

Bridges play an important security role, discouraging sniffing by restricting Ethernet frames to the collision domains between the actual

2. The limit for 10-Mbps Ethernets is four repeaters between any two MAUs in one collision domain. Using three or more repeaters imposes additional limitations on segment lengths and media types. The 100-Mbps Ethernets are limited to one Class I or two Class II repeaters per collision domain, also with additional restrictions on segment lengths. Gigabit Ethernet (1,000-Mbps) introduces a new "buffered distributor" hub type that is not subject to the collision domain restrictions.

sender and receiver. It is also possible to define a filter for an individual port on some bridges, restricting the kind of Ethernet frames that will be received or transmitted.

An Ethernet switch (sometimes called a switching hub) is basically a multiport bridge. If a single device is connected to a port on a switching hub, there is no contention for bandwidth on the network between that device and the hub; throughput up to the full media rate is limited only by the capacity of the switch. A switch can be a good way to deal with an overloaded network, and as a bridge it avoids the eavesdropping problem. Some switches are hybrid devices that include repeater hubs and IP routing features.

Routers are like bridges in that they selectively echo traffic onto different interfaces, but they do so in terms of IP addresses and packets instead of MAC addresses and Ethernet frames. Thus, routers are not really just Ethernet devices—a single router can offer a variety of different types of ports, including Ethernet and V.35 synchronous connections.

The important distinction here is that routers can control the propagation of Ethernet broadcast traffic, which bridges do not filter very well. A broadcast frame is sent to an Ethernet MAC address of all 1s and is received by every interface on the network. This may be done for a variety of reasons, but with IP it is most commonly used as part of the ARP. Routers can compartmentalize broadcast and other traffic into different IP networks or subnets, passing packets on to different interfaces only when absolutely necessary.

Full-duplex link segments

Light or electrical impulses can only travel but so fast, and if competing Ethernet senders must detect a collision within a certain period of time then the laws of physics place hard limits on the distance across a collision domain. This becomes more of a problem as data rates increase. Fortunately, a way has been found to sidestep collision domain limitations: full-duplex link segments.

A link segment connects just two devices, one at each end. If neither device is a repeater (i.e., both are either a computer, bridge, or router), and there are two different conductors for signals between them, then both devices can transmit simultaneously without concern for collisions, and the connection between them is said to be a full-duplex link segment. This makes it possible to run Ethernet over two strands of fiber optic cable for 20 km or more.

The different types of Ethernet

There are quite a number of different types of Ethernet; only those you are most likely to encounter are summarized here. This is not a complete list of all the specifications—if you need to design an Ethernet network, use an authoritative reference such as Spurgeon's *Ethernet Configuration Guidelines* [7]. A beginning community network is not likely to need bandwidth beyond the 10 Mbps Ethernet types today, but when making cabling investments it is a good idea to plan for future, higher-speed links.

- **10BASE-2 (thin-net):**

  Access type: Mixing (more than two MAUs can be attached to one segment);

  Media: RG58 A/U or C/U 50-ohm coaxial cable (approximately 3/16 in thick);

  Speed: 10 Mbps;

  Connectors: Bayonet Neill-Concelman (BNC) tee coax connectors;

  Distance limit: 185m (606.9 ft) per segment;

  Other limitations: No more than 4 cm (1.57 in) between MAU interface and the main cable; no less than 0.5m (19.6 in) between MAU attachments to the main cable; no more than 30 MAU attachments to a single segment; up to four repeaters between any two MAUs in a collision domain;

  Comments: Convenient and inexpensive for connecting a few machines in a small area. Bad connections can be exasperatingly difficult to find and fix. Problems with one NIC, connection, or cable can easily take down the entire network. Relatively easy for users to eavesdrop on each other's traffic. Not a good choice to run between buildings because of problems with electrical ground differentials. Not recommended for anything more than very small scale use.

■ 10BASE-T (10-Mbps twisted pair):

Access type: Link (only two MAUs attached to one segment);

Media: Four wires (two each to send and to receive), minimum Electronics Industry Association/Telecommunications Industry Association (EIA/TIA) Category 3 unshielded twisted pair (voice-grade UTP); Cat-5 data-grade UTP recommended (see comments);

Speed: 10 Mbps;

Connectors: 8-pin RJ-45 style connectors (similar to modular telephone plug);

Distance limit: 100m (328 ft) per segment;

Other limitations: No more than 90m (295 ft) recommended from hub in wiring closet to outlet plate in user area, allowing for 15 ft or so of patch cable from the outlet to the computer, signal losses at connectors, etc.; up to four repeaters between any two MAUs in a collision domain;

Comments: 10BASE-T is the current technology of choice for large-scale installations, but will likely be displaced by 100BASE-TX as bandwidth demands increase. Each individual computer is connected to its own 10BASE-T link running back to a hub or switch (multiport bridge) in a wiring closet. Problems with one connection usually do not affect the others. Connecting users with switches or hubs with security features can prevent eavesdropping. 10BASE-T was designed to run over existing voice-grade telephone wiring (Category 3 UTP), but if you use Cat-5 or better UTP you may be able to upgrade without rewiring to 100 Mbps 100BASE-TX, and possibly even to 1000BASE-T. If you do plan to upgrade, consider using enhanced Cat-5 (Cat-5E) or Cat-6 cable, and make sure you conform to the installation and testing standards ANSI/TIA/EIA-568A, ANSI/TIA/EIA-TSB-67, and ANSI/TIA/EIA-TSB-95. As with all copper-media Ethernet, 10BASE-T is not a good choice to run between buildings because of problems with electrical ground differentials.

■ 10BASE-FL (10-Mbps fiber optic):

Access type: Link (only two MAUs attached to one segment);

Media: Two strands (one to send, the other to receive) of fiber optic cable, most often 62.5/125 multimode fiber (MMF) with a 62.5-micron core and a 125-micron outer cladding, driven with an 850-nm wavelength light source. Some vendors offer media converters that can be used to run 10BASE-FL full-duplex links (i.e., between bridges, routers, or computers) over longer distances with single-mode fiber (SMF), typically with an 8 to 10-micron core and a 125-micron cladding (e.g. 10/125);

Speed: 10 Mbps;

Connectors: Traditionally ST-type[3] fiber connectors, also known as International Standards Organization/International Electrotechnical Committee (ISO/IEC) International Standard Bayonet Fiber Optic Connector (BFOC)/2.5. SC-type connectors are becomming increasingly popular, but are more frequently found on higher-speed devices;

Distance limit: 2,000m (6,561 ft) per segment with MMF, 20 km (about 12.4 miles) or more possible on full-duplex links over SMF;

Other limitations: For MMF, no more than 12.5 dB of optical signal loss per segment; budget 5 dB per 1,000m and 0.5–2.0 dB per connector junction, depending on the quality of the connection. Up to four repeaters between any two MAUs in a collision domain but with reduced cable lengths;

Comments: Like all fiber links, avoids any interbuilding electrical ground differential problems. The same 62.5/125-MMF cable used for 10-Mbps 10BASE-FL can later be used with 100-Mbps 100BASE-FX or ATM at even higher data rates. 1,000-Mbps Ethernet is specified to run over 62.5/125 MMF as well, but for shorter distances (see 1000BASE-SX, 1000BASE-LX). If any older fiber optic inter-repeater link (FOIRL) components

---

3. ST is a registered trademark of AT&T.

are used, the maximum 10BASE-FL distance drops to 1,000m. Glue-on connectors have been found to be more reliable than the crimp-on type, but the quality of the latter is improving.

- **100BASE-TX (100-Mbps twisted pair):**

Access type: Link (only two MAUs attached to one segment);

Media: Four wires (two each to send and to receive), minimum EIA/TIA Category 5 unshielded twisted pair (data-grade UTP); enhanced Cat-5 (Cat-5E) or Cat-6 will be less likely to have problems;

Speed: 100 Mbps;

Connectors: 8-pin RJ-45 style connectors (same as 10BASE-T);

Distance limit: 100m (328 ft) per segment;

Other limitations: No more than two 100m segments connected by one repeater in a single collision domain. No more than 90m (295 ft) recommended from the hub in wiring closet to the outlet plate in user area, allowing for 15 ft or so of patch cable from the outlet to the computer and signal losses at connectors. Maximum of one class I or two class II repeaters per collision domain;

Comments: Many 100BASE-TX network interface cards can automatically sense and adapt to either 10BASE-T or 100BASE-TX signaling rates. The 100BASE-TX specification was written for Cat-5 cable, but some sites have had significant problems upgrading their 10BASE-T Cat-5 installations [8]. If you do plan to upgrade, consider using enhanced Cat-5 (Cat-5E) or Cat-6 cable, and make sure you conform to the installation and testing standards ANSI/TIA/EIA-568A, ANSI/TIA/EIA-TSB-67, and ANSI/TIA/EIA-TSB-95. If you envision a near-term commitment to very high bandwidth applications (e.g., medical imaging), it may be cost effective to switch now to a fiber-optic cabling strategy (see 100BASE-SX). As with all copper-media Ethernet, 100BASE-TX is not a good choice to run between buildings because of problems with electrical ground differentials.

■ 100BASE-FX (100-Mbps fiber optic):

Access type: Link (only two MAUs attached to one segment);

Media: Two strands (one to send, the other to receive) of fiber optic cable, most often MMF with a 62.5-micron core and a 125-micron outer cladding (62.5/125 MMF). Some vendors offer media converters that can be used to run 100BASE-FL full-duplex links (i.e. between bridges, routers, or computers) over longer distances with SMF, typically with an 8 to 10-micron core and a 125-micron cladding (e.g., 10/125).

Speed: 100 Mbps;

Connectors: Duplex SC connectors (recommended), or M keyed FDDI media interface connectors (MICs), or ST connectors (same as for 10BASE-FL);

Distance limit: With MMF: 412m (1,351 ft), less if repeaters are used within a collision domain; 2000m (6,561 ft) over full-duplex links; with SMF: 20 km (about 12.4 miles) or more;

Other limitations: For MMF: No more than 11 dB of optical signal loss per segment; budget 2 dB per 1,000m and 0.5 to 2.0 dB per connector junction, depending on the quality of the connection. Maximum of one class I or two class II repeaters per collision domain;

Comments: Immune to interbuilding electrical ground differential problems. 62.5/125 MMF cabling put in place today for 10BASE-FL can later be used for 100BASE-FX full-duplex links. 1000-Mbps Ethernet is specified to run over 62.5/125 MMF as well, but for shorter distances (see 1000BASE-SX, 1000BASE-LX);

■ 100BASE-SX (100-Mbps short wavelength fiber optic— draft standard):

Access type: Link (only two MAUs attached to one segment);

Media: Two strands (one to send, the other to receive) of 62.5/125 MMF cable driven with an 850-nm wavelength light source;

Speed: 100 Mbps;

Connectors: SC or ST connectors;

Distance limit: 300m (984 ft);

Other limitations: Must be a full-duplex link segment—only non-repeater devices may be connected (computers, bridges, or routers);

Comments: Although as of early 1999 100BASE-SX is still only a draft standard (proposed by the Short Wavelength Fast Ethernet Alliance), some products are already available. The intent is to use short-wavelength LEDs to reduce network interface costs to about half that of 100BASE-FX devices, thus promoting the use of fiber-optic cabling in short-range applications such as connecting user workstations to a switch. This development, along with new lower-cost fiber connectors, may soon bring the cost of fiber Ethernet installations down to the same level as copper [9]. Like all fiber-based Ethernet options, 100BASE-SX is immune to inter-building electrical ground differential problems. MMF cable runs for 100BASE-SX can later be upgraded to gigabit Ethernet, but distance is limited to 275m for some types (see 1000BASE-SX, 1000BASE-LX).

- **1000BASE-T (1,000-Mbps twisted pair—draft standard):**

Access type: Link (only two MAUs attached to one segment);

Media: Eight wires (four each to send and to receive), minimum EIA/TIA Category 5 unshielded twisted pair (data-grade UTP). enhanced Cat-5 (Cat-5e) or Cat-6 recommended;

Speed: 1,000 Mbps;

Connectors: 8-pin RJ-45 style connectors (same as for 10BASE-T and 100BASE-TX, but all eight wires used instead of just four);

Distance limit: 100m (328 ft) per segment;

Other limitations: Cabling must be installed in conformance with the installation and testing standards ANSI/TIA/EIA-568A, ANSI/TIA/EIA-TSB-67, and ANSI/TIA/EIA-TSB-95; no more than 90m (295 ft) recommended from switch in wiring closet to outlet plate in user area, allowing for 15 ft or so of patch cable from the outlet to the computer and signal losses at connectors;

Comments: The 1000BASE-T specification is scheduled for final release in early 1999. Members of the IEEE 1000BASE-T task force have stated that "... any link that is currently using 100BASE-TX should easily support 1000BASE-T" [10]. Unfortunately, some sites have had significant problems upgrading their 10BASE-T Cat-5 installations to 100BASE-TX [8]. Enhanced Cat-5 and Cat-6 specifications are in development. If you envision a near-term commitment to very high bandwidth applications (e.g., medical imaging), it may be cost effective to switch now to a fiber-optic cabling strategy (see 1000BASE-SX) that would offer fewer problems when you later upgrade to 1,000 Mbps or beyond. As with all copper-media Ethernet, 1000BASE-T is not a good choice to run between buildings because of problems with electrical ground differentials.

- **1000BASE-SX (1000-Mbps short wavelength fiber optic):**

Access type: Link (only two MAUs attached to one segment);

Media: Two strands (one to send, the other to receive) of MMF optic cable, either 62.5/125-micron, or 50/125-micron, driven with an 850-nm wavelength light source;

Speed: 1,000 Mbps;

Connectors: Duplex SC connectors;

Distance limit: For full-duplex link segments, 275m over 62.5/125 MMF (200 MHz·km modal bandwidth measured at 850 nm), or 550m over 50/125 MMF;

Comments: Immune to interbuilding electrical ground differential problems. The shortwave light source makes 1000BASE-SX devices less expensive that the 1,300-nm wavelength 1000BASE-LX variety, but the latter is needed for longer distances. Most existing MMF installations in the United States use 62.5/125-micron fiber, and most 10BASE-FL and 100BASE-FX products are designed for 62.5/125 MMF. Some vendors promote switching to 50/125-micron MMF for new installations because at higher data rates it can perform over significantly longer distances.

■ **1000BASE-LX (1000-Mbps long wavelength fiber optic):**

Access type: Link (only two MAUs attached to one segment);

Media: Two strands (one to send, the other to receive) of fiber optic cable, either 62.5/125-micron multimode, 50/125-micron multimode, or 10/125-micron singlemode, driven with a 1,300-nm wavelength light source;

Speed: 1000 Mbps;

Connectors: Duplex SC connectors;

Distance limit: For full-duplex link segments, 440m over 62.5/125 MMF (500 MHz·km modal bandwidth measured at 1,300 nm), 550m over 50/125 MMF, or 5km over 10/125 SMF;

Comments: Immune to interbuilding electrical ground differential problems. The longwave light source makes 1000BASE-LX devices more expensive than the 850-nm wavelength 1000BASE-SX devices, but the former is needed for longer distances. Most existing MMF installations use 62.5/125-micron fiber, and most 10BASE-FL and 100BASE-FX products are designed for 62.5/125 MMF. Some vendors promote switching to 50/125-micron MMF for new installations because at higher data rates it can perform over significantly longer distances.

■ **1000BASE-CX (1000-Mbps Twinax):**

Access type: Link (only two MAUs attached to one segment);

Media: Balanced, shielded, 150-ohm two-pair coaxial cable;

Speed: 1000 Mbps;

Distance limit: 25m;

Comments: This technically conservative standard was designed to be released quickly in order to allow equipment supporting 1000BASE-SX and 1000BASE-LX to be inexpensively connected over short distances. It will likely be supplanted by 1000BASE-T jumpers in the future. As with all copper-media Ethernet, 1000BASE-CX is not a good choice to run between buildings because of problems with electrical ground differentials.

Token ring

Token ring is like Ethernet in that it can be used to carry IP or other protocols' packets over short to medium distances at a relatively high speed, 16 Mbps. New token ring network equipment, however, tends to be very expensive compared to Ethernet equipment. A typical token ring card for a PC today costs more than $200, while very high-quality 100-Mbps Ethernet cards are now available for under $75. Because of Ethernet's overwhelming popularity in the marketplace and the onset of new networking technologies such as 100BASE-TX, this situation is not likely to change in token ring's favor.

The appropriate choice is clear if a new network is being installed. If a token-ring network already exists, the decision is less obvious. You certainly want to avoid investing in additional token-ring equipment if at all possible. The short-term solution of bridging IP traffic between token-ring and Ethernet networks tends to be messy and problematic. Making a complete switch to Ethernet may cost more up front but it will result in savings over time as the network expands, and the costs of buying and maintaining a bridge are avoided.

ATM

Asynchronous transfer mode (ATM) is a new type of networking technology that can provide high-speed connections over UTP and fiber optic cabling. It is designed to carry data, voice, and video simultaneously. Quality of service (QOS) designations can be established for individual ATM connections to ensure that traffic that absolutely, positively, has to get there in the next 15 milliseconds will make it in time, avoiding dropouts in voice or video continuity. Another new ATM feature is the ability to switch IP traffic across an entire ATM network as if it was a single router hop—this can speed IP packet transmissions and avoid overloading routers on a congested network, an increasing concern as data rates accelerate and routing tables grow longer. Standards are in place and products exist for running ATM at 155 Mbps for 2,000m over the same

62.5/125 MMF used for 10BASE-FL Ethernet, and 15km or more over singlemode fiber.

The drawbacks to ATM are cost and complexity. Integrating ATM into an IP network for anything more than a WAN link can be a daunting task, and an ATM switch represents an additional piece of hardware to purchase, install, configure, and maintain in addition to the Ethernet hubs, switches, and routers you will also inevitably need. New protocols are being developed that may succeed in adding QOS features to IP without the use of ATM.

Gigabit Ethernet represents the greatest challenge to ATM at this point. If all you need is raw bandwidth for local IP data traffic, 1,000 Mbps can deliver it, if your router can keep up! However, if you need to seamlessly and reliably carry voice, video, and data over the same network, especially in a WAN context, then you will need ATM's capabilities.

Wireless point-to-point links

Transporting data from one location to another with radio signals is not new, but recent developments have made it a more attractive option in the community network context. New technology, including spread spectrum radio, has made wireless data communications less expensive, more secure, more resistant to interference, and able to run over longer distances at higher speeds. Equipment costs are still high in relation to traditional telephone line modems for delivering data to a scattered population of individual users, but when compared to leased-line costs wireless point-to-point links can be an attractive way to stitch together distant sites into a community network.

There are many wireless products [11]. Some require special licenses, but many operate in the unlicensed industrial, scientific, and medical (ISM) bands: 900 MHz (902-928 MHz), 2.4 GHz (2400-2483.5 MHz), and 5.8 GHz (5725-5850 MHz). These frequency ranges have been set aside in the United States for unlicensed low-power spread spectrum communications; other countries have designated ISM spectrum as well but may use different frequencies. Although there is a recently approved standard for radio LANs (IEEE 802.11), products from different vendors often will not interoperate, and the capabilities of ISM products vary widely. Vendors currently claim speeds ranging up to 25 Mbps, and distances up to 50 km. A typical device that delivers 2 Mbps over 20 km might run $5,000 per end including the antenna. That is an expensive up-front cost, but it becomes very attractive when compared to leased line charges of $500 per month, or fiber cable installation of $5–10 per foot.

Unfortunately, establishing an effective wireless link and keeping it running reliably is not as easy as it might sound. The most important consideration is the location of the end stations. ISM equipment is designed to run over line-of-sight paths. If you can clearly see one antenna from the other today, so far so good—but will it always be that way? More than one wireless link has been obscured by growing trees or a newly constructed building.

Interference is a normal part of the territory in radio communications. Heavy rain or snow, microwave ovens, and other users' equipment can all be sources of interference in the ISM bands. Spread spectrum devices automatically adapt to these problems, but it comes at a cost—all that error correction and packet retransmission saps bandwidth. Vendors usually advertise the signalling rate for their products. Actual data throughput can often be only 50–70% of the signalling rate, so be sure to properly size a wireless link for your real bandwidth requirements.[4]

The most insidious wireless problem is unknown interference that intermittently knocks out a previously reliable link. You may need to hire an engineer just to locate the source of the trouble. If it's someone legitimately operating equipment that's also designed to run in the ISM bands, there's not much you can do other than to try to reconfigure your equipment to work around the problem—that's one of the liabilities of using unlicensed frequencies. There may be alternative wireless data services available in your area based on licensed spectrum, such as local multipoint distribution service (LMDS).

Despite the problems, wireless has come a long way in the last three years, and future developments are likely to make it an even more attractive strategy for linking together a community network, especially when distances are large.

---

4. Data throughput over Ethernet, by comparison, can reach as much as 80–95% of the signalling rate.

Application-level protocols: HTTP, Gopher, FTP, SMTP, POP, RFC822, IMAP, LDAP, NNTP, and Telnet

Internet application-level protocols correspond most closely to the actual tools people use to accomplish real tasks on the network. Application clients sit at the top of the protocol stack, employing the services of all the lower layers to carry on a conversation in their own unique languages (the application protocols) with a remote server. For most applications, the lower-level stack is TCP/IP/Ethernet for users with direct connections and TCP/IP/SLIP or TCP/IP/PPP for modem-connected users. The following is a brief list of the most common applications and application protocols:

- Hypertext transport protocol (HTTP): HTTP is used by WWW clients (browsers) to send requests to Web servers, which then return Web pages for display. Most contemporary Web browsers are multifunctional; they also provide access to Gopher, FTP, e-mail, and other services that are based on their own protocols instead of HTTP.

- Gopher: Gopher clients perform a similar function to Web clients, providing users with a point-and-click interface to peruse and retrieve information offered by a remote Gopher server. Gopher data, however, are presented textually in structured hierarchical menus instead of free-form Web pages.

- File transfer protocol (FTP): FTP is used by an FTP client to copy files to or from a remote computer running an FTP server. Web server machines often also run an FTP server to allow web designers to update their pages.

- Simple mail transfer protocol (SMTP): SMTP is used to send e-mail files between Internet-connected computers. An e-mail client starts an SMTP connection with a remote e-mail server (typically a UNIX server application known as sendmail) and passes it a message along with sender and recipient addresses. The format of the message itself is defined by RFC822.

- Post office protocol (POP): Mail can be delivered to a server computer that holds it in a mailbox for later delivery to a client application. Typically, the e-mail client uses POP to retrieve the message from the server. Two variants of POP are in common use: POP-2 and POP-3. (Note that POP is also an acronym for "point of presence," a location with Internet connectivity.)

- Internet message access protocol (IMAP): IMAP is an alternative to POP for retrieving mail from a server and extends POP's capabilities to allow users to manage their messages by permanently storing them in different folders within their server mailbox. This means that users can access old as well as new mail from any computer with an IMAP client, but it also means that the server must have sufficient storage capacity to hold all the saved messages for every user.

- Lightweight directory access protocol (LDAP): LDAP is a protocol for searching and updating online directory information. Online directories can not only hold information about individuals and organizations but also encryption keys, providing a foundation for secure communications.

- Network news transfer protocol (NNTP): NNTP is used to transport network news articles between news servers and between news servers and news readers.

- Telnet: Telnet is a protocol that can be used to conduct a text-based interaction with a remote computer, the same as if you were typing commands directly on that computer's console. There are other protocols, most notably X Windows, that can be used to conduct a graphically based interaction with a remote computer.

Bringing the Internet to your town

We've talked at length about different tools that can be used to establish high-speed Internet access. Now let us consider how some of those options might be applied to support a community network.

Bringing in Internet access is referred to as "establishing a point of presence," a place where Internet connectivity is present and hence a location where other networks can be connected to the Internet. There are basically three parts to establishing a local point of presence: the

Internet feed, the access line that delivers it, and the equipment needed to connect to the access line. Each of these items has an associated cost, although you may end up paying a single fee to one vendor for a bundle of services and equipment.

Getting an Internet feed

Your Internet access provider (IAP) is your connection to the rest of the Internet. The provider either has a direct connection to an interconnect point where your packets can be exchanged with those of other IAPs and their customers, or can pass packets to an interconnect point through another access provider, perhaps a national service provider (NSP). What you buy from the IAP is the IP packet exchange service, not the wire that carries the packets.

The distinctions between the different types of providers are confusing. An Internet service provider (ISP) can provide a variety of Internet services, including e-mail or a server that hosts Web pages. An IAP provides Internet access, services such as a modem pool or leased-line connections you can use to communicate over the Internet. Some ISPs also provide IAP-type services. An NSP is a special type of IAP, a large-scale Internet-access wholesaler marketing at a national or international level, often providing Internet access to smaller providers, which in turn sell that access to individual businesses and consumers.

The block of IP addresses you use for your network connections are usually assigned by the IAP. You will need to obtain a network domain such as "bev.net" or "blacksburg.va.us" to serve as the basis for the domain addresses to which these IP addresses will correspond. You can e-mail the form to apply for your domain yourself (see the subsection "Domain name service" for site addresses), or your IAP can do it for you. If you are establishing a domain name identity for your community, make sure someone in your organization is listed as the administrative contact. Domain ownership could become a legal issue if your IAP is the only one listed on the form and you decide to switch to a different provider.

One option for establishing an Internet feed is to get a direct link to an NSP. This may provide you with faster service and fewer problems, but it may also be more expensive. Another possibility is to establish a link with a local IAP that has its own NSP link and remarkets pieces of it to others.

A third option is to find a business or organization in the community that already has an Internet link and explore with them the possibility of adding a general town feed to that service.

It definitely will be worth your while to consider all the options. The monthly cost will vary depending on where you are located and the data rate of your connection to the provider, but typical charges for T1-level access can run upward of $2,000 per month in addition to a startup fee—and that is just for IP packet service; the access line itself is a separate cost.

Selecting a provider

Selecting the IAP (or NSP) to supply the Internet feed can be a complex decision. Part of the equation is the bottom line, but the provider's reliability also needs to be considered. Is the IAP increasing capacity to keep pace with growth? Will the IAP be buying any additional equipment to handle the increased demands your access will place on its network? What kind of track record does the IAP have resolving problems for existing customers? The alt.internet.services newsgroup is one place to look for customers of an IAP under consideration. The Internet also offers a number of IAP lists and directories; try searching Yahoo! for IAPs [12].

Exactly what will your IAP provide?

When comparing IAPs, consider the following issues:

- Are there limits on the number of IP addresses the community network can have? Will it cost more to get more? What if the network wants to sell access to businesses around town instead of just providing connections to the local schools and library?

- Will the IAP operate a primary or secondary DNS for the network? If so, is there a limit on the number of addresses the IAP will handle? Running a name server is not a severe load for a reasonable number of addresses, but the network will need to have at least two separate name server machines acting as the source of authority for your domain.

- Will the IAP provide other services such as a network news feed, e-mail boxes, or space on an FTP or Web server?

- Will the IAP also provide a router or CSU/DSU to terminate the leased line? Some providers sell or lease equipment on an optional basis; some may require that you purchase your equipment from them; and some may even require that you purchase the equipment at the IAP's end of the leased line as well! It is not uncommon for an IAP to specify which brands and models of CSU/DSU and router it will support.

- Some IAPs will negotiate and contract with the phone company for a community network's leased access line as well as sell the actual Internet feed. Be sure to compare the IAP's package deal to what it would cost to purchase the Internet feed and the leased line separately.

- Terms of agreements with IAPs vary. Watch out for long-term contracts! Costs are falling, and the variety of different connection options is increasing all the time. The three-year contract that looks like such a good deal today may be a real loser in as little as 18 months. Bear in mind, however, that switching providers will incur not only the startup fee for the new provider but also the phone company fee for switching over the access line.

The access line

A community that wants to establish an Internet network utility should plan on getting at least a 1.5-Mbps T1 leased line to carry the Internet feed. In real-world terms, a T1 is generally accepted as having sufficient bandwidth to carry the traffic generated by about 200 to 300 average 28,800-bps modem users [13]. Multiplying 28,800 by 200 gives 5.76 Mbps, almost four times the actual 1.536-Mbps data capacity of a T1, but it is not likely that all those modem users will press the enter key at the same time. T1 speed is not only necessary to handle a significant number of users but is also needed to realize the potential of 10-Mbps Ethernet connections when facilities such as libraries, schools, and businesses are direct-wired.

There will be servers running at the point of presence and elsewhere in the community, so there is also a need for the full-time connection characteristic of a leased line. Intermittent, slow-to-establish connections such as those provided by modems over circuit-switched lines are not appropriate for networks running servers. Packets requesting a Web page cannot wait while a connection is dialed up.

T1 prices are falling and vary significantly between localities. In the Blacksburg area, typical costs are now under $500 per month with a comparable setup charge. The cost is higher if the distance is long since T1 leased lines are priced by the mile.

What are the alternatives? In the past, 56-Kbps leased lines were frequently used, but by contemporary standards they are barely sufficient to support a few modem users. ISDN BRI service is generally switched, subject to a per-minute usage charge, and runs up to only about 128 Kbps. Asymmetric cable modems and DSL can deliver data downstream to the community network at T1 rates or better, but the upstream performance is not sufficient for servers or large numbers of users. All of these choices have the added drawback of requiring an investment in equipment that cannot be used if there is a later upgrade to a T1.

There are a few other options with data rates that may be sufficient in situations in which the community needs to start small and work its way up. Fractional T1, which supplies some subset of the T1 rate in a multiple of 64 Kbps, say, 384 Kbps or 768 Kbps, may cost somewhat less to lease than a full T1 line. One of the DSL variants that supports an upstream data rate comparable to the downstream rate may also be available at fractional T1 speeds. An IAP should also charge less for an Internet feed at a slower rate, since less internal capacity will be needed to handle the traffic. If the right fractional T1 equipment is selected, it can also be used later to upgrade to a full T1.

Frame relay is another intermediate-bandwidth possibility, if it is available. Like fractional T1, properly chosen equipment can be used later for a full T1 link. The lack of mileage charges makes frame relay particularly attractive if the IAP is a long distance away.

Equipment needed for an Internet leased line

Once a provider has been lined up and a line leased from the phone company, what equipment is needed at the community's end? The answer is basically the same for T1, fractional T1, and frame relay: a CSU/DSU that connects to the T1 line and a two-port router that has an Ethernet interface for the network on one side and a V.35 synchronous connection for the CSU/DSU on the other (see Figure 10.2).

Routers vary tremendously in capabilities and corresponding costs, but in most situations they represent a nonredundant point of potential network failure, so it is worthwhile to invest in a good one. Almost all routers that can handle full T1 connections can handle fractional T1 as well, but the reverse may not be true. Some routers include frame relay capability or can be configured with that as an option. Prices have fallen 20–30% over the last three years; by the spring of 1999 a basic, very high-quality two-port router listed for around $1,700. Less expensive models (with lesser capabilities) can be had for $1,200 or so, and the IAP that provides the network feed may offer high-quality routers at a substantial discount, especially when purchased in quantity.

A well-chosen CSU/DSU can be used with frame relay as well as fractional and full T1. CSU/DSUs also represent a single point of failure but tend to be less complicated than routers and therefore less vulnerable to problems. A good manageable CSU/DSU capable of handling the full T1 rate runs around $1,000. Many routers now offer WAN link interfaces with integrated CSU/DSUs.

The IAP may offer guidelines about what brand of equipment should be used to terminate the leased line; some will even require specific models of CSU/DSU or router to simplify their network management. Whether or not the IAP requires particular equipment, it is a good idea to pay careful attention to its recommendations.

A 10BASE-T hub is necessary to make the Ethernet connectivity provided by the router available to the various computers that will actually use it. Most hubs act as simple repeaters; they echo every packet that appears on any port out all the other ports. A basic 8-port dumb 10BASE-T hub can cost as little as $30; a 24-port stackable hub with both 10BASE-T and 100BASE-TX capability, remote management, and security features can run $1,000 or more.

One of the nice things about Ethernet networking equipment is that as the network grows and new higher-capacity equipment is moved in, the older, lower-capacity equipment can still be used out on the new network frontiers. That dumb 8-port hub may not be appropriate for the center of the network forever, but there will always be a need for it somewhere.

The CSU/DSU and router should have RS232 serial ports that can be used for configuration management. You will need some kind of PC with terminal emulation software such as HyperTerminal (included with Windows95) or Kermit to deal with this. A laptop is handy for that kind of thing, as well as for diagnosing other problems.

The first cold winter night that the network goes down, the network administrator will want to be able to get to those management ports from home, so it makes sense to invest in a modem and a serial line switch that allows him or her to select which device to talk to. (Note that you won't need the serial line switch if the CSU/DSU is built into the router.) The modem's phone line at the point of presence is also handy for contacting vendors' technical support about recalcitrant equipment.

A miscellaneous but important item is power conditioning. Beyond a simple surge protector, a properly sized uninterruptable power supply (UPS) that can carry your equipment for 10 minutes or so will avoid many network problems precipitated by momentary power glitches.

Where does all this stuff go?

This question might seem trivial, but it deserves serious thought. The obvious answer is to start out with the Internet feed located in the first building that is to be supplied with high-speed Ethernet access. Doing that avoids additional costs incurred by piping Ethernet from the community Internet point of presence to users in a different building.

If possible, the Internet feed should be located centrally in relation to all the sites that are to be supplied with direct Ethernet connections: libraries, schools, businesses, government offices, community centers. High-speed communications links between buildings can be accomplished with inexpensive multi-mode fiber (MMF) technology, but it is only possible to do it without intermediate equipment when the cable length does not exceed 2,000m (about 6,500 ft). For example, if two primary targets for Ethernet capability, say, the town library and an office park, are two miles apart, consider locating the Internet feed at a point halfway between them. If longer cable runs are required, single-mode fiber, SMF, can be used, but the transceivers on each end of the cable will be at least two to three times as expensive. Also keep in mind that the installed cost for fiber cabling can run $5–8/ft above ground and $10/ft below ground.

Other physical considerations for the point-of-presence location are security and a relatively clean, temperature- and humidity-controlled environment.[5] If the equipment is going to be set up in an unheated area, the manufacturer should be consulted to make sure it can tolerate such conditions. A large amount of floor space is not necessary; 50 sq ft should be sufficient to start.

Local delivery

The T1 is in, the brand-new router and CSU/DSU are hooked up with an 8-port hub, and you have just plugged in your laptop. That T1 sure is fast when only one person is using it!

But what about everybody else in town? You can hook up six more people on the hub (the router took one port, and your laptop is on the other), but 10BASE-T can only run 100m (about 328 ft), and it's not a good idea to run nonfiber links between buildings because of electrical grounding problems. The library is a half-mile away, and what about dialing in from home?

There are a variety of ways to distribute Internet access around town once a network point of presence has been established. At least two basic needs have to be addressed: (1) high speed with a constant connection and (2) ubiquitous access. The answer to the first need is to install some kind of dedicated links running from the point of presence to the users that need dedicated access, and the answer to the second is a modem pool.

The modem pool

Ubiquitous access is a requirement, and a modem pool is the only answer. Even if every office in town is wired with Ethernet, the high-tech

---

5. One BEV town Ethernet failure was due to lint accumulation on a router located in a laundry room. Another router was stolen, then abandoned when the thief realized it was not a VCR.

workaholics among us will still need to get online on weekends to download the latest software off the Net. People will need to dial in when they are on the road, and there will also be schools and other sites for which the expense of a faster connection will not be initially justifiable. One way or another (we will consider some options later), a modem pool with Internet access is a requirement.

The structure of an Internet connection was outlined in the section "Basic Internet technology." Now let's follow a scenario of a user dialing in so we can see in detail what a modem pool actually involves.

John, back in Grays Harbor, is at it again. Dinner is over, and he's bored with TV, so he decides to see what's happening on the Net. He starts his computer and double-clicks on his Web browser application. The browser program starts and internally signals the network driver program on his computer to start a connection. Part of the driver (called a dialer) sends a special configuration command to the modem (typed manually it would look something like AT E1 L2 M1 Q0 V1 X4 &D0 &C1), then it sends another command with the modem pool phone number to actually make the call (ATDT1234567).

The modem pool phone number is associated with a collection of phone lines—a rotary—that are each in turn connected to an individual modem at the Internet point of presence (see Figure 10.3). The RS232 serial data connection on the other end of each modem is connected to an individual port on a device called a terminal server. A terminal server is a special-purpose computer that provides a serial port user with access to a high-speed network. (Modern modem pools often use *access servers*, which integrate the modems and terminal server into a single device, but the funtions remain distinct.)

The call from John's computer is answered by the modem the rotary selects for him. John's modem and the modem at the point of presence then briefly hiss at each other to decide how fast they can talk. Once the introductions are complete, the modems settle down to quietly passing data back and forth between John's computer and the terminal server.

As soon as the terminal server ascertains that the modems have connected successfully, it makes its own challenge: "Username:" or something of that sort. On reading that request as it arrives from the modem, the dialer program sends back the username with which it was configured

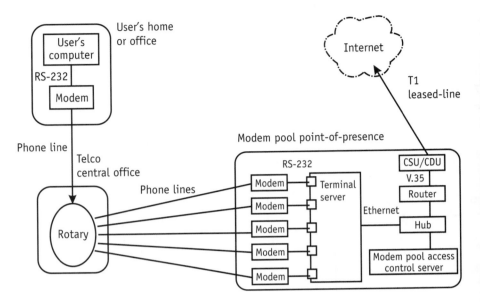

Figure 10.3   Modem pool equipment.

and then prompts John to type in a secret password to authenticate his call. Other data may also be exchanged to further describe exactly what type of connection John's network driver wants to start and what address John's computer will have on the network.

Once these formalities have been dispensed with, both the terminal server and John's network driver shift into what might be called packet transport mode, in which they use a special computer communications language to carry network packets over the serial connection. When a modem connection is used to access the Internet, the protocol used almost certainly will be SLIP, CSLIP, or PPP.

As long as the SLIP (or PPP) connection is up, John's computer is effectively on the Internet at the point of presence where the modem pool is located. Packets sent from John's machine travel out onto the Internet at large, and packets destined for his machine are funneled back through the terminal server, point-of-presence modem, and phone line, back through John's modem to the network driver program, and ultimately to the Web browser on his computer.

Data milking machine: modem pool equipment

A hundred phone lines, 100 modems, 100 terminal server ports, and connections between all of them. If that sounds complicated, there is a reason: Modem pools are one of the most complex aspects of providing Internet access. Nevertheless, it can be done. The problem is to do it right, so that users' connections are reliable and busy signals infrequent.

To begin, consider the modems themselves. Faced with the prospect of having to purchase 10 or more modems at once, you might be inclined to pick up a recent PC hardware catalog and shop for bargains. That would be a bad idea—modems are notoriously cantankerous, particularly when they are communicating with one made by a different manufacturer. You cannot control what kind of modem your users will buy (actually, you know what your users will buy: the cheapest mongrel modems they can get). Your best defense is to buy modems that have a strong reputation for compatibility with all other brands. Each high-quality standalone modem will probably cost $100–200.

How many modems are needed? That largely depends on the online behavior of the network's users, but for small systems one modem for every 10 registered users is a popular ratio. Thus, an initial setup with 16 modems and phone lines should be able to handle a user population of about 160, as long as users do not develop a tendency to keep their connections up even when they are only using it to check their empty e-mail boxes every 60 seconds. Such behavior can be limited by setting the terminal server to drop connections after a maximum time is reached.

If a network is running more than 10 modems and the administrator does not want to have to go in and power-cycle them occasionally to resolve a midnight problem, an access server system that integrates the modems and terminal server into a single chassis should be seriously considered. Such a system can provide modem management, the ability to reconfigure the modems when they get wedged into some incontrovertible state. A good modem management system will try to resolve problems on the fly; if it cannot, it will "busy out" the problematic phone line so no user will get stuck with a dead modem.

The good news is that access servers can be used to build reliable pools with many hundreds of ports and prices are falling. The bad news is that at current prices a high-quality access server with 48 modems can still cost $300–500 per line.

Purchasing smaller access servers tends to increase the per-port cost significantly. You can try to cut corners by buying a multiport board that plugs into a PC running Windows NT or UNIX and use that to support dialup connections as well as some other part of the operation. However, that solution tends to drag down the server's performance and will not scale well for larger numbers of ports.

Other modem details

If your community network embarks on the journey that is running a modem pool, there are a number of other details to contend with:

- *User client software:* The users of a modem pool will need to run Internet-capable software on their computers, and not everyone will have a new machine that came with network clients already installed. Will they obtain this software from the community network or purchase it somewhere else? A community network will have to handle multiple platforms, at least Windows and Mac. You can try to put together a conglomeration of freeware/ shareware software and hand that out on diskettes or CDs, but that means someone will have to assemble the packages. License issues can be tricky, especially for noneducational institutions. One popular option is for the network to arrange a volume purchase agreement with a vendor that sells such a package and resell it at a discount to customers.

- *User support:* Using someone else's software is particularly attractive when you consider the support issue—customers can call the software vendor's support line instead of the network's. However, even if the software issue is farmed out, the network will still have to provide some degree of support to address the problems of users trying to dial into the modem pool with their mongrel modems. That means employing intelligent, resourceful, patient, untiring, resilient, customer-oriented people to answer phones and provide friendly assistance ("I forgot my password again!") for at least eight hours most weekdays. Each of these folks will need a phone and at least one computer, preferably one for each platform variant that the network supports. The latter need can be finessed somewhat by using one monitor switched between different system units or by using a Mac that has an internal Intel board running Windows.

■ *Billing and accounting:* The network will need income to support all its hardware, and that means billing the users somehow. Along with billing come all the usual retail business issues of handling cash and dealing with credit cards and bad checks. The minimum requirement will be a fairly sophisticated database on a PC. Employing a banded usage charging scheme instead of a flat rate (e.g., first 30 hours of connect time for $15, $1 per hour thereafter) will make careful tracking and logging of each user's modem pool activity a much more critical issue and require a commensurate investment in system resources. Note that some vendors' terminal or access server equipment will offer better accounting features than others.

The pros and cons of a modem pool

There is a positive side to running a modem pool—it can be a paying proposition. Just 32 lines supporting 320 registered users paying $20 per month would generate $6,400 a month, or $76,800 a year. The user population is likely to increase, and if the network is already running a modem pool, adding extra-charge services like Web page development for businesses would be trivial.

The problem is that running a modem pool is a business in itself. If the purpose of the network is to use new networking technology to increase communication within the community and to provide an environment that enables high-tech economic development, your plate is already full. Farming the modem pool out to someone else is the best way to deal with this.

With any luck, a modem pool is already up and running in your area. If not, one in a neighboring calling area may be interested in installing foreign exchange (FX) lines to provide callers in your community with a local dialing number.

If you are fortunate enough to have a local modem pool IAP, try to work with them; a T1 should be able to handle at least 200 modem lines.

If the IAP is not running that many, they should be looking for additional users to help defray the $2,500 a month or so being spent on an Internet feed from their access provider and a leased line from the phone company. The community network can help a local modem pool IAP use that bandwidth, by adding businesses, schools, and libraries with Ethernet connections.

High-speed local delivery

Currently, only modem pools can deliver Internet connectivity to everyone anywhere, but at locations with large concentrations of users or significant need, another option can provide something better: truly high-speed, full-time Ethernet connections.

A modem pool connection is established by a user on demand, then dropped when it is no longer needed, freeing up the phone line and equipment at the modem pool end for another user. Dedicated leased-line, frame relay, wireless, or symmetric DSL connections not only carry data at rates 10–100 times faster than a modem, they also provide full-time access, which is required to support servers providing Web page, e-mail, and other services to the Internet at large.

Consider what happens when John's packets from Grays Harbor arrive to request the BEV Web page. They cannot wait for a modem pool connection to be established, and the already slow speed of a modem line would grind to a crawl when more than one remote user requested service at the same time.

The latter concern also highlights the issue of providing Internet access to large groups of users in schools, libraries, or businesses—they cannot practically share a single modem and modem connection. For the cost of providing a modem, a phone line, and modem pool access for each of them, a much better option could be had, once a network point of presence has been established in town.

Another growing issue is the increasing bandwidth demand of Internet applications. Over the last few years, Gopher and later the Web increased bandwidth needs for an individual user by a factor of 5 or 10. They did that by popularizing the transmission of images, which require many more bytes to represent them than simple textual information. We are already seeing an increasing exchange of audio and short video clips, and it recently has become possible to do real-time videoconferencing with relatively inexpensive equipment directly over the Internet if you have Ethernet and a leased-line network feed. These applications will ultimately increase bandwidth requirements by another order of magnitude or more; users in the next few years will come to expect and demand such capabilities.

Local high-speed candidates

Locations that can effectively use local high-speed access have at least one of two criteria: significant numbers of users in a single location, or users that need high data rates and / or full-time connections for their particular application. Some examples follow.

- *Schools and libraries:* These sites have large numbers of users and increasing bandwidth requirements. A library with a full-time Internet connection can run its own server to allow off-site access to its catalog and is a logical focal point for community information on the network.

- *Local government offices:* These are important sites to connect when they have a significant concentration of users at one location. Small across-town offices with two or three occasional users may be more economically served with modem pool, ISDN, DSL, or cable modem connections. Planning departments and other groups that use mapping software may need high-bandwidth connections to support the transfer of images between locations.

- *Businesses:* Large businesses in a community may already have Internet connections, but attaching to a locally operated point of presence may be an attractive economic alternative if a reliable connection can be ensured. Geographically scattered small businesses with few users are difficult to connect in a piecemeal fashion, but a number of different firms grouped in one office building are an excellent opportunity. Installing high-speed connections in an office complex is a powerful way to encourage high-tech business development.

- *Apartment complexes:* An apartment complex is not an obvious target at the outset, but housing with Ethernet access in a community with a high-tech worker or student population makes sense. Also, an apartment complex with Ethernet is an instant high-tech business incubator: An ISP starting with one server computer in an apartment connected to the Internet can market Web services to clients anywhere in the world. That may seem preposterous, but it has been happening in Blacksburg since 1993. By 1997 the BEV apartment Ethernet network had become a net producer of information on the Internet, transmitting more data than it received.

Local high-speed structure: linked LANs

An Ethernet network running in a building or a series of connected buildings constitutes a LAN. When local Ethernet networks are connected to a community's point of presence, that community network acts as an IAP, similar to the way the network's IAP delivers an Internet feed. As with the community's connection to its IAP, a number of different technology options are available for connecting local LANs to the point of presence.

The most obvious connection is a 1.5-Mbps T1 line. Adding a connection of that type requires a router port and a CSU/DSU at the point of presence, a leased line from the phone company, and a CSU/DSU, router, and hub at the remote site. New, so-called branch office router offerings may integrate several of these functions into a single box.

T1-level service will likely be more expensive than other options, but it may be required to handle sites with large numbers of users or that need to use high-bandwidth applications such as video over the network. Figure 10.4 illustrates the use of T1 and fractional T1 for local high-speed distribution.

A line to deliver fractional T1, data service over a T1-type line but at a lower rate, such as 256 or 512 Kbps, may be somewhat less expensive to lease from the phone company. Keep in mind, however, that a 256-Kbps line is roughly equivalent to only nine 28.8-Kbps modems, which will not be satisfactory for a site with 40 simultaneously used computers. Purchasing a CSU/DSU or router that cannot handle the full T1 rate may be less expensive but, because bandwidth requirements are likely to increase, is not advisable.

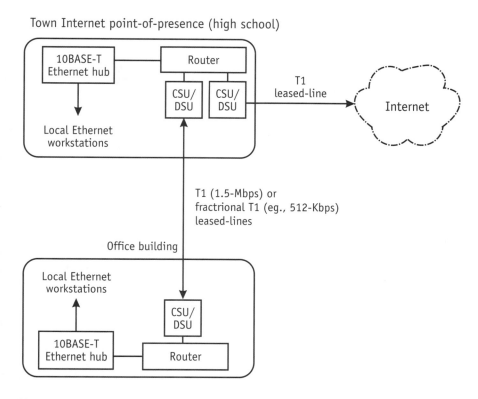

Town Internet point-of-presence (high school)

Figure 10.4   Local high-speed distribution—T1 or fractional T1.

As a more economical alternative than T1, frame relay can be an attractive option. Distance charges do not usually apply to frame relay connections, and a single router port and CSU/DSU at the point of presence can support PVCs to several different user sites. If your community network intends to take this route, make sure the frame relay router at the point of presence can support as many PVCs as are likely to be needed. Figure 10.5 illustrates frame relay used for local high-speed distribution.

The limitations of frame relay may preclude its use in certain situations. The CIR is typically only half the peak rate, so a PVC with a peak rate of 1.024 Mbps is guaranteed only 512 Kbps. That rate may not be sufficient to service sites with many users or sustained high-bandwidth application needs such as video. Thirty heavy users of a PVC with a CIR/burst

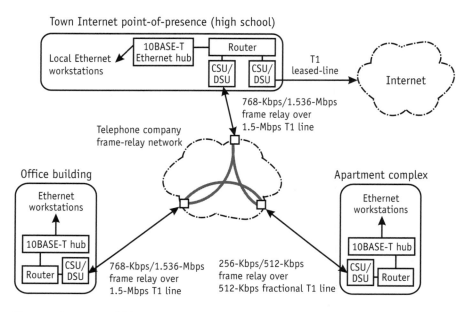

Figure 10.5   Local high-speed distribution—frame relay.

rates of 256/512 Kbps would not see much better response than each would using a 28.8 Kbps modem.

In general, frame relay would be best used for far-flung sites with moderate numbers of users and bandwidth requirements. A single 1.5-Mbps access line and router port at the central site should be able to provide full-time connectivity for five to ten such locations at CIR/burst rates of 512/1024 Kbps to 256/512 Kbps.

The Montgomery-Floyd Regional Library has used frame relay to extend its Ethernet network beyond the initial T1-based access in Blacksburg to two additional branches in Christiansburg and Floyd. A full-mesh network with permanent virtual circuits connecting each site to the other two is supported by Bay Networks BayStack Access Node routers and ADC/Kentrox DataSMART CSU/DSUs in each location. Frame relay service purchased from Bell Atlantic avoids the high mileage costs that would be charged for traditional T1 leased-line service between these widely separated locations. Figure 10.6 illustrates this network.

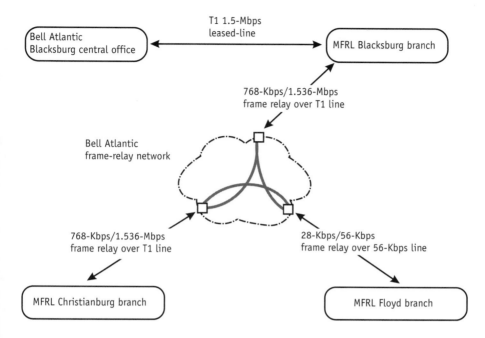

Figure 10.6   Montgomery-Floyd Regional Library T1 and frame relay network.

Local information superhighway: 10-Mbps Ethernet links

Some situations may call for faster local service than what T1 links can offer, for example, a large number of users at a local site that heavily access a server running at the community network point of presence or at a different local site. 10BASE-FL fiber links can deliver a full 10 Mbps to support those needs. Figure 10.7 illustrates use of a 10BASE-FL link for local high-speed distribution.

One scenario to be aware of is when a substantial number of users access the network through one point of presence, but a significant proportion of their activity focuses on servers run at a different point of presence. If the only path between the users and the servers runs along the T1 lines linking the two points of presence to the Internet, much of that expensive leased-line bandwidth will be absorbed by that activity. This is roughly equivalent to using a long-distance service to call your neighbor down the street. A situation like that calls for a 10-Mbps Ethernet short

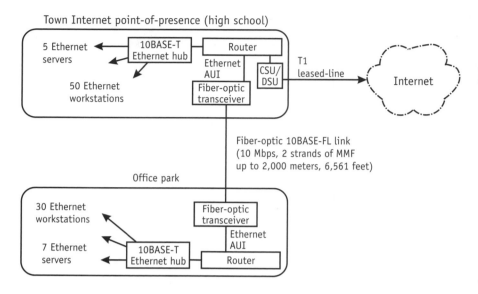

Town Internet point-of-presence (high school)

Figure 10.7   Local high-speed distribution—10BASE-FL link.

cut between the local points of presence: a fiber link running up to 2,000m (6,561 ft) between router Ethernet interfaces using relatively inexpensive 10BASE-FL fiber optic transceivers (FOTs), also known as fiber optic MAUs (FOMAUs).

An FOT attaches to a standard Ethernet MAU interface on one side and has a pair of ST connectors for 10BASE-FL standard fiber optic cabling (one strand to transmit, the other to receive) on the other. 10BASE-FL components use MMF with a 62.5-micron core and 125-micron outer cladding (MMF 62.5/125); some vendors' equipment accommodates 50/125 micron cable as well.

That sounds too good to be true: a full 10 Mbps for over a mile, and instead of a $1,000 CSU/DSU at each end, all that is needed is a $140 FOT. The catch is getting fiber between the sites that need service. The local phone company may be willing to lease so-called dark fiber, strands of a cable it already has in place or will install. Since it will be used instead of a more traditional leased-line service the phone company could sell you, their dark fiber may be quite expensive.

Other options are to lease fiber strands from some other large communications service user or vendor that happens to have fiber in your community—railroads, power companies, cable TV, and long-distance telephone companies are all candidates—or install your own. Actual costs will vary depending on the situation, but contractors in the southwest Virginia area are currently installing fiber for $5–8/ft above ground and $10/ft for interducted underground fiber.

Router and bandwidth planning

How does a community network handle all these network satellites being put into orbit? Adding another CSU/DSU and router every time another site (or three, in the case of frame relay) hooks up gets expensive, not to mention crowded when all the equipment starts stacking up.

One alternative is to purchase a multiport router, which is like a basic two-port router but with additional synchronous (V.35 for T1 CSU/DSUs) or Ethernet interfaces. Some models provide a fixed number of ports; others offer a modular setup in which ports can be added by buying additional cards or stackable units from the vendor. In general, a network that has a multiport router with card slots available should be able to add ports for $500 to $750 per synchronous line, plus $1,000 for the CSU/DSU to attach between it and the T1 line. (Some router interface cards feature an integrated CSU/DSU that will reduce the total cost.) Figure 10.8 illustrates use of a multiport router.

Other local high-speed possibilities

Other possibilities for establishing local network links at higher-than-modem speeds include ISDN BRI, cable modems, DSL, and point-to-point wireless connections.

ISDN BRI service can provide a 128-Kbps connection over a switched circuit. An ISDN bridge or router can be configured to start a connection only when packets actually need to flow over the link. That is important, because per-minute charges often apply to ISDN calls. Check with your local phone company for rates and availability. The 128-Kbps ISDN BRI rate makes it unsuitable to support large groups or active Internet servers, but a classroom of 15 lightly used computers can work well through an ISDN BRI Ethernet bridge.

Cable modems and DSL are promising technologies but not yet universally available. Both are usually asymmetric with regard to their

Town Internet point-of-presence (high school)

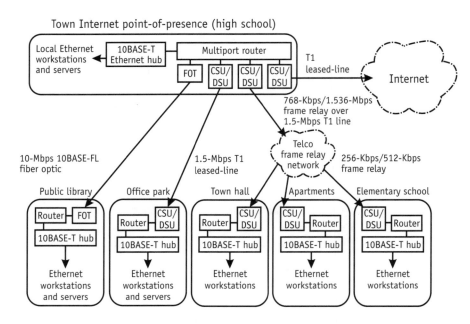

Figure 10.8   Local high-speed distribution—multiport router.

data delivery: Data can be received by the end user faster (perhaps as fast as 2 Mbps) than it can be sent (perhaps only a few 100 Kbps). Cable modems have the advantage of a higher theoretical speed (up to 30 Mbps), but also suffer from being a shared-access medium: As the number of active users sharing one data channel grows, the bandwidth available to each of them shrinks. It might be possible to use a service of this type to support a remote classroom or small library, but it would not be appropriate to locate an active Internet server or a large number of users at the end of an asymmetric link.

Some types of DSL connections are symmetric and can deliver data rates in both directions at speeds comparable to fractional T1 lines. This is an option worth exploring, if it is available in your area. Be aware, however, that the copper phone line delivering your DSL service will be maintained by the local phone company, and their repair policy for such connections may be different than it is for T1s. Some users have reported long delays getting DSL line problems fixed.

Point-to-point wireless links are targets of opportunity and necessity. You have the opportunity to use wireless when you have good line-of-sight antenna locations at the sites to be connected, unobscured by present or future trees or buildings. You may by necessity have to resort to wireless if the month-to-month cost of the alternatives is more than your budget can bear. A wireless link can easily cost $5,000 per end once antenna costs are included, but that can beat $500 per month for a leased T1 in less than 2 years.

Serving the public: required and optional Internet servers

In addition to the communications equipment described in the previous sections, there are a number of network server applications a community network will either have to run or may choose to run. Unlike the specialized hardware devices described up to this point, these services run on one or more general-purpose computers attached to the network, so there will also be choices to make in terms of the server hardware and operating system.

Required services directly support the function of the network. Two such services are the modem pool authorization service and the modem pool billing system. The first service keeps a list of userids and passwords for everyone with modem pool access. When a user dials the modem pool and enters a userid and password, the terminal server consults the authorization server to see if they are correct. The billing system tracks and logs user activity reported by the terminal server.

Some terminal or access servers may provide internal support for these functions, but more commonly they are implemented as software running on a separate computer. Before a terminal server is purchased, its access control and access logging mechanisms should be carefully investigated, since they may set unexpected requirements in terms of servers and operating systems.

Even if a community network is not running a modem pool, it may still be useful to run an authorization service to support other services with identified users, such as e-mail or some Web applications.

The most critical service on the Internet

The most critical required service is DNS, which keeps track of which IP address (e.g., 198.82.165.36) corresponds to a given fully qualified domain name (e.g., www.bev.net) and vice versa.

There will be IP addresses and domain names corresponding to them for every interface attached to the network. Two different machines will have to provide authoritative DNS for those addresses. Your IAP may be able to provide both primary and secondary service, but that is not likely unless the community network has a relatively small number of addresses. It is also inadvisable for a network's only name service to run at the other end of the Internet feed. Virtually every network application begins every action with a flurry of DNS lookups, and that traffic should be kept off the easily congested IAP access line as much as possible. The community network should plan to operate at least a primary name server, assuming the IAP will operate a secondary one.

Other services: the good, the bad, and the ugly

Most DNS programs are relatively well-behaved as Internet services go. They do not require an exorbitant amount of memory or processor resources, as long as they do not maintain thousands of addresses.

Authorization services vary greatly in their system requirements, depending on how they are implemented and how many users are registered. If you expect to register hundreds of users, look for authorization software backed by a database that can retrieve records based on a number of different indexes (e.g., by userid, by last name and first name, or by id number).

Beyond an authorization service and DNS, there are a number of other Internet services a community network may choose to operate. The most obvious is a WWW server, which is virtually a requirement for a community network. The Web is the easiest, most universally accessible way to provide information to a user community (e.g., that the system will be down next Tuesday so the router can be upgraded).

Web service can be hosted by quite lightweight systems, provided that the hit rate (the number of requests per minute) is low and the amount of data being returned per hit is not large. Page designers should avoid bandwidth hogs such as large images and video clips on frequently accessed pages. There should not be a problem in terms of performance running reasonably sized DNS and Web services on the same machine, although it may be wise to segregate DNS to its own machine for security reasons.

The next most user-desired service is probably e-mail. An e-mail server can grow to become a major resource hog, but that is not likely to be a problem for up to a few hundred users. Part of the problem lies in the frequency with which users check to see if the server has new mail for them. Some server implementations are more efficient than others in terms of how they send and receive messages and how they provide information to users about the messages in their mailboxes. E-mail service for 500 or more users calls for server software with efficiency features.

One angle on e-mail services is to have someone else provide them. E-mail generally requires less bandwidth than the Web, for instance, because most e-mail traffic is text. That means that it will be less of a problem for users to access e-mail services somewhere beyond the community network point of presence. Modem pool operators almost always operate an e-mail server; someone else in the community providing modem pool service may be interested in selling e-mail-only accounts to Ethernet-connected users.

If a Web server is being run as a community resource, an FTP server on that same system will allow people in the community to post information on it. The more the maintenance of information on the Web server can be distributed to volunteers and authorities in the community, the richer and more frequently updated the information will be. FTP allows those individuals to directly replace files on the Web server without requiring the intervention of a system administrator. Items can be submitted by e-mail for centralized posting, but this will quickly become a drain on the community network staff as the breadth and depth of information on the server increases.

FTP requires some form of authorization and user identification. It is essential that the operating system supporting the Web and FTP servers be capable of allowing only certain users to update specific files. It is also desirable for the operating system to implement quotas, that is, limitations on the amount of disk space different individuals can use.

Now for the most truly unmanageable network service of them all: network news. This is not a reflection on network news content but on server behavior—a news server is capable of grinding almost any machine down to an absolute crawl. A full news feed from the IAP (many supply one, for reasons that will become obvious shortly) can clog your Internet feed with as much data as a T1 can carry; much of that volume is not text but encoded images. This flood can be throttled to some extent by being selective about which news groups you choose to download.

The alternative to a community network providing news service is for the news aficionados among its users to independently go out and subscribe to news servers elsewhere on the Internet, then pull articles in over your T1 connection as they read them. If a large number of users start reading the same articles, it may be possible to save bandwidth in the long run by simply maintaining a server with the articles users want at the local community network point of presence. That is why IAPs often offer a news feed: It preserves bandwidth on their connection to the Internet backbone if they pull in news traffic only once, instead of having each of their customers independently pull a separate feed.

As with e-mail, network news is an excellent service to find someone else to provide locally. Many modem pool providers run news servers. That activity is probably best operated by a private business; it may be difficult for local government or facilities such as libraries or schools to offer news service due to controversy over content issues.

How much iron do you need?

What server hardware (and, to some extent, what operating system) you need to run depends on what services your community network intends to provide.

The only services a small-scale network needs to offer are primary DNS and some small- to medium-scale Web service with a bit of FTP traffic to maintain the pages. Those services could be easily supported with a single, small-scale machine such as a 200-MHz PowerMac or a 200-MHz Intel Pentium running Windows NT or Linux (a UNIX variant for Intel hardware). A more mainstream (and more expensive) UNIX solution would be a Compaq Alpha Server DS10, or a Sun Ultra 5. Memory and disk space needs will vary depending on the operating system, but 64 MB of RAM and 2 GB of disk would be the minimum requirements. There are now also turnkey "Internet server appliance" products available that bundle hardware and software into an easily-managed system designed for small businesses—these may serve as a good starting point for a community network as well.

For a medium-scale scenario, in addition to the preceding requirements, you need to support e-mail and modem pool authorization and billing. Authorization goes on the Web server machine (extra RAM may be necessary to avoid virtual memory disk swapping delays), and e-mail goes on an additional machine of at least the scale described in the above paragraph. Modem access logging goes onto both machines in duplicate files. The grunt work of preparing modem pool bills and keeping track of user addresses and such can be delegated to a frequently backed-up Mac or PC running a good database package. With two machines, a network can run a local secondary DNS (that will keep second-try DNS requests off your Internet feed when users' first attempts find the Web server busy), and it will be more convenient to test new software or recover from a system failure.

A large-scale scenario would include running a news server. That actually does not require exceedingly powerful hardware—the above-described CPUs are adequate—but to accept a full feed of news coming in per day requires lots of disk and possibly more RAM depending on how many users are expected to request news. News could be run adequately on the same machine as the e-mail server, especially if the news-reading population is not large or you do not accept a full feed, but watch out! News tends to expand its system utilization and absorb all available resources. A news server is best isolated to its own hardware, perhaps sharing that machine with only a secondary DNS server.

The great O/S debate

A discussion of servers would not be complete without adding fuel to that great traditional bonfire, "Which operating system is the best?" All systems have assets; in general, whichever one gets the job done with the least fuss and bother is the best one for your network. That operating system usually will be the one the network administrator and staff are the most familiar with, assuming reliable and reasonably priced implementations of the required server software are available. That being said, here are some subjective comments regarding server operating systems and hardware.

- *UNIX:* The Internet was invented using UNIX. Most popular Internet application servers are available in source code form for UNIX for free on the Internet and also in the form of compiled binary executables for popular UNIX platforms. UNIX runs on just about every kind of significant-scale hardware; of particular note is Linux, the most popular of a flock of open-source UNIX variants. The BEV has found DEC (now Compaq) Alphas running Digital UNIX to be reliable, easy to find software for, and an excellent bargain in terms of price and performance.

  People unfamiliar with UNIX, even those with extensive experience in other operating systems, find the UNIX learning curve daunting. Linux is much easier for novices to deal with than some of the more traditional types of UNIX, but in any form UNIX is complex. If Microsoft Windows is a minivan with an automatic transmission, UNIX is an 18-wheeler with 15 forward gears and a three-speed rear axle. The truck comparison is appropriate: UNIX has evolved over many years to carry a wide variety of loads. But if you're going to run UNIX, you either need to learn how to drive the truck or you need to hire a truck driver.

- *Microsoft Windows NT:* Windows NT has emerged as the principal challenge to UNIX for heavy-duty server applications. I have spoken with ISPs happily running all their services on NT machines. That notwithstanding, Linux seems to have recently overtaken NT in the small-scale ISP realm. If no UNIX expertise is available to your network, Windows NT is worth considering. Note that Windows NT runs not only on Intel hardware but also on Compaq Alphas.

- *Microsoft Windows95 and Windows98:* Some Internet server applications for Windows95 and Windows98 can be found, but generally they are not robust enough to satisfy the needs of a community ISP.

- *Macintosh:* Mac Web servers are said by some to be the easiest to manage on the Net and often have integrated FTP servers. Other server applications such as DNS and e-mail can be found for the Mac as well, but they are less widely used. Macs have an excellent track record in resisting hacker attacks, a major system administrator headache and time-waster. One alternative worth considering would be to put a number of Mac Web servers in schools and community centers, where information can be easily updated, and carry links to those servers on a central community-wide Web server.

In 1999 Apple released a new Unix-based operating system for the Mac. If this successfully grafts traditional Mac ease-of-use onto cast-iron Unix internals, Macs could become much more widely used as Internet servers in the next few years.

In summary, if the community network staff has some experience running UNIX, it is best to run UNIX. If your staff is more inclined toward a different operating environment, find someone successfully running the needed Internet servers on that platform and pattern your system after their system. Reliable and robust software, operating systems, and hardware are the foundation needed to provide reliable and robust services to the network's users.

Evolution of the Blacksburg town Internet

When the BEV began in 1993, Internet access was supplied by the only available local provider: Virginia Tech. BEV users signed up for access to the Virginia Tech modem pool, and with Bell Atlantic's assistance Ethernet networks were installed in local schools, libraries, museums, and apartment complexes. The network path between all these users and BEV's servers at Virginia Tech was direct and quick, because all of them were tied to the Internet through Virginia Tech.

By the spring of 1996, tides of change were sweeping the BEV in new directions. Many of those forces were the fruit of the BEV's continued cultivation of Internet use in the community.

Two new private modem pools were in operation, responding to the market created in part by the BEV project, and BEV users not affiliated with the university were asked to move off the Virginia Tech modem pool by July 1, 1996. Both those providers had their own Internet feed. One, NRVnet (http://www.nrv.net/; NRV stands for New River Valley), was based in Blacksburg and supported access through a local Blacksburg phone number as well as phone exchanges in three surrounding counties. The other, Citizens InterNET Service (http://www.swva.net/), was based in Floyd County, Virginia, south of Blacksburg, and supported access with local numbers in nine different southwest Virginia counties. Additional modem pool access providers quickly moved into the area.

Modem pools were only one avenue of growth. An increasing number of apartment building and commercial property owners installed Ethernet wiring and inquired about how to connect their buildings to the Internet. That increase was in part driven by the large population of students actively using Ethernet in on-campus housing; about 1,900 dorm rooms had Ethernet by the fall of 1996, and all were wired by 1998. As those students moved off campus, they were not likely to be satisfied with mere modem connections. Waiting lists grew for Ethernet-equipped apartments.

In addition, many new businesses started up in the town to provide a variety of Internet-related services [14]. Many of these ventures outgrew their small Ethernet-connected apartment beginnings and sought more mainstream office space with Ethernet in town or at the Virginia Tech Corporate Research Center.

By 1998 local Internet access provider options had developed to the point that it was no longer necessary for Virginia Tech to continue to support Ethernet connectivity to the apartment complexes. These networks were transferred to private IAPs.

As network use has grown in Blacksburg, the migration of different segments to private providers has been an appropriate outcome, and a necessary one as the number of users surpassed the university's ability to manage them. The creation of new business opportunities was one of the project objectives from the outset. This diversification, however, has presented some new problems.

When their network connection ran directly to Virginia Tech, users had been accustomed to a reliable, high-speed connection between their home computer and campus facilities, including the BEV servers. The IAPs that now supply their connectivity have Internet feeds that are separate from Virginia Tech's. Since the only link between them and Virginia Tech is via their respective access providers (possibly even two different NSPs and a congested national network access point), town Internet users accessing on-campus services often experience Internet backbone network delays. This traffic pattern also increases the load on Virginia Tech's Internet feed as well as that of the local provider.

Local network access point

The appropriate response to these concerns is to assemble a local network access point (NAP). A NAP is a kind of demilitarized free-trade zone where different IAPs can meet to exchange packets. In the center of the NAP stands a simple Ethernet switch. Clustered around the switch like little castles (actually, mounted in a rack next to it and connected to the Ethernet ports) are small routers, each belonging to a different IAP. Each router guards the path back into that IAP's realm, its own network, which lies at the far end of the T1 or Frame Relay connection the IAP has installed to the NAP premises. Figure 10.9 illustrates this arrangement.

Each IAP can make peering agreements with other IAPs that are connected to the NAP. If users at IAP #1 want to look at data that happens to be on a server attached to IAP #2's network, their request packets can flow from IAP #1's network up their T1 line and through IAP #1's router to the NAP Ethernet switch. From the switch they flow into IAP #2's router, down their T1 line, and back to IAP #2's network where the server is located. Response packets follow the same route, but in reverse. This may seem like a long and complicated journey, but in fact it's quite short—a two-router-hop path is a huge shortcut compared to the previous connection through separate national service providers and a national-level NAP, which can amount to 15 router hops or more.

It might seem simpler to connect multiple IAPs to a single router, but that would make the router, and more significantly the management of that router, a centralized resource. The strategy of using a low-level Ethernet switch as the medium of exchange at the NAP minimizes costs, and places the responsibility for maintaining the appropriate routing tables and sufficient router horsepower directly on the parties with the most direct interest, the connecting IAPs.

The benefits of the NAP will not only be savings in IAP Internet feed bandwidth, fewer e-mail delays, and faster surfing on local Web servers. Quicker and more reliable packet exchange between local users will also make it possible to explore innovative new high-bandwidth services, such as interactive audio and video links.

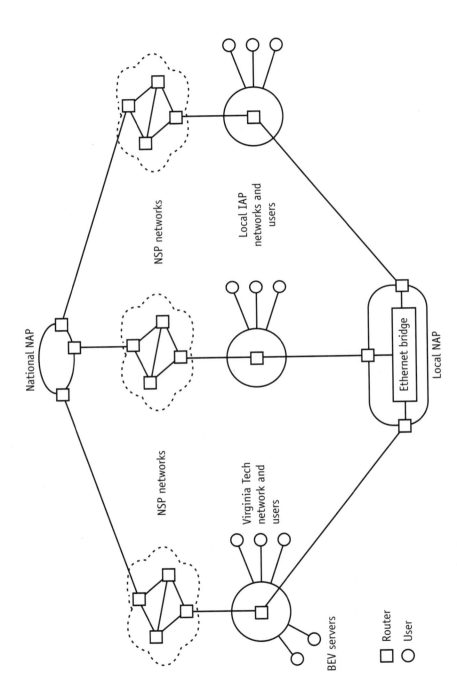

National NAP

NSP networks

Local IAP
networks and
users

NSP networks

Virginia Tech
network and
users

Ethernet bridge

Local NAP

BEV servers

□ Router
○ User

Figure 10.9 Local NAP (network access point).

Fiber on main street: new technology opportunities

Once an Internet point of presence has been established in a town, the problem of extending Ethernet access to other sites presents itself. This can be a formidable hurdle to cross at the outset, because of the complexity and costs associated with links based on traditional WAN technology such as T1 leased lines and multiport routers.

New options have emerged in the marketplace in the form of fiber optic cabling and Ethernet switches. This technology can be applied in an evolutionary manner to gradually extend direct Ethernet access to an increasing number of locations within about a mile or so of a point of presence, establishing in the process a fiber optic infrastructure that could be used in the future to support much higher data rates. This option is particularly attractive for small towns, where the distances between community-oriented sites often are short.

What fiber can do, now and in the future

Readily accessible and relatively inexpensive 10BASE-FL components can deliver 10-Mbps Ethernet connectivity today at a range of up to 2,000m (6,561 ft) using standard multi-mode fiber (MMF) terminated with inexpensive ST connectors. A $140 FOT can be attached to an MAU interface on any normal Ethernet hub, bridge, or switch. Longer 10-Mbps runs can be supported with single-mode fiber (SMF), but the transceiver hardware will be at least two to three times as expensive. Expensive multiport routers can be avoided in the beginning phases—an early implementation of a town network is unlikely to generate sufficient traffic to overwhelm a simple bridged configuration.

Cable installation costs are significant but not out of the question when considered as a long-term investment: $5–8/ft for above ground and about $10/ft for in-ground ducted fiber. Follow-on installation of additional fiber in existing ducts is very reasonable—only about $1–2/ft. Fiber terminations are in the range of $10–25 each, including the connector. Pole right of way may be leased in the neighborhood of $25 per pole or may be subject to a one-time fee.

In the future, the same MMF cabling put in place to support 10-Mbps 10BASE-FL links can be used to support 100 Mbps or even higher speeds.

That means that a fiber optic infrastructure put into place today will still have value in tomorrow's networks, in which voice, data, and video will share the same communications path.

Phase 1: establishing the point of presence

The starting point for a small-town Internet utility is the local Internet point of presence. The point of presence should be established in a central location relative to other sites likely to need Internet access, in order to reduce fiber cabling costs. A less critical but significant criterion for the point of presence site is the number of potential users at that specific location. Establishing the point of presence at the town library with floor space for 10 or more computer workstations, for instance, would mean that you could start teaching hands-on Internet classes and provide walk-in access to residents almost as soon as the T1 was in place, without waiting for the installation of fiber cabling to other sites.

Phase 2: bridging to the first few sites

Fiber connections from the point of presence to other sites initially can be connected to an Ethernet multiport bridge (an Ethernet switch) instead of a router. Bridges manage packets based on lower-level MAC addresses, while routers are more sophisticated, using IP or other higher-level protocol addresses. For a beginning network with relatively low traffic levels, bridges will serve quite adequately. This is a significant economic advantage, since multiport routers may cost more than $1,000 per port. A 12-port switch with reasonable capabilities can be obtained for under $75 per port.

Figure 10.10 illustrates the second phase of the town network evolution, in which one port of a bridge with three AUI ports is connected to the Ethernet network at the point of presence. An FOT is connected to another of the AUI ports, and two strands of MMF terminated with ST connectors are attached to send/receive ports on the FOT.

At the other end of the MMF run, up to 2,000m away, the two strands are connected to a second FOT, which is in turn attached to a 10BASE-T hub. Twisted-pair Ethernet cabling radiates from the hub to workstations up to 100m away (about 300 ft).

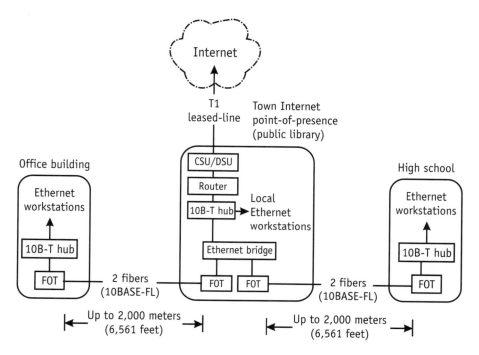

Figure 10.10　Fiber on Main Street phase 2—simple bridged network.

Phase 3: bridging to a group of sites at a distance

In some cases, a group of different sites needing Internet connectivity may be located near each other but are a long distance from the central point of presence. In that case, a single fiber run can be established from the point of presence to a distribution point centrally located relative to the group of remote sites, another bridge placed there, and then fiber installed connecting that bridge to hubs in the actual buildings to be serviced. Figure 10.11 shows a network with a secondary bridge.

This may require the construction of an equipment hut in which the secondary bridge is to reside. Another less expensive option is to locate the bridge in a secure location on the premises of one of the sites to be served.

Phase 4: central routing, more bridging

Eventually traffic volume will increase at some sites to the point that the bridged-network structure can no longer adequately carry the

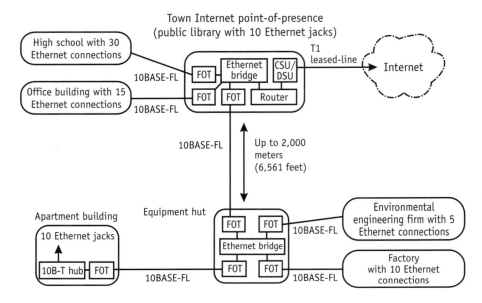

Figure 10.11   Fiber on Main Street phase 3—network with secondary bridge.

load. That will happen for some legs of the network before it happens on others. By the time it does, there should be sufficient users on those segments to support the cost of the multiport router required to meet their needs.

As those high-volume network links move to router ports, the bridge ports they previously occupied can then be used for new low-volume connections, preserving the investment in the equipment. This scenario is typical for an evolving network based on nonproprietary communications standards: As traffic increases and more sophisticated equipment is used near the core, the older, less sophisticated equipment can be pushed out toward the periphery to serve less demanding feeder links.

Potential problems: right-of-way, IP address planning, security, non-IP protocols

This somewhat blithe presentation of a fiber-based community network has ignored a number of details, none of which is a show stopper, but a few are worthy of particular note.

■ *Right-of-way:* Who owns the utility poles? What about the easement for in-ground utility installation? Current utility providers seeking to go into the data distribution business have an advantage here, but the new telecommunications law mandates equal access to distribution facilities.

■ *Future address planning:* A bridged system without many users will quite happily function regardless of how IP addresses are assigned. At some point, however, routers will start appearing at various points in the network. When that happens, it will suddenly become necessary for IP addresses to be assigned in some logical fashion. For instance, site A uses IP addresses 198.82.212.32-63, site B uses 64–95, and so forth. Exactly how those addresses should be allocated is a complex art called subnetting. Suffice it to say that a fledging community network should be sure to get someone to help plan the IP address assignments to avoid the need to reassign everyone's address when the system starts using routers in the future.

■ *Security:* Businesses that link their Ethernet LAN into the Internet through a bridged community network will not be pleased if students at the local high school (or students on another continent) find a way into their corporate files. The router that terminates the T1 line at the point of presence can be configured to limit certain types of access to certain local IP addresses from the Internet at large, but it will not control access between machines on the bridged network. The bottom line is that businesses that have serious security concerns can purchase their own firewall router to terminate the fiber run on their premises and should investigate an application gateway firewall as well.

■ *Non-IP protocols:* Ethernet can be used to transport a variety of protocols besides IP. Some that are frequently used in business LANs are Novell's IPX, Microsoft's NetBEUI, and Appletalk. These protocols are sometimes referred to as "chatty," that is, they tend to use a lot of broadcast packets and can quickly congest a network as usage increases. For that reason, and because of security concerns, bridge port filters should be set up to pass only IP traffic. A community network will probably be pressed by some users to pass a non-IP protocol between selected sites across town. The best option to support that would be for those users to purchase routing hardware that specifically supports the transport of those protocols across a WAN.

Technology planning summary

This section is a summary of some of the options presented in this chapter. Keep in mind that the real community network is the user's view of the information. The technology planner's job is to make choices about how to support that vision, not to find some way to use every possible tool.

Services

Which network-based information services does your community network need?

- Local Web pages;
- Local chat service;
- E-mail lists;
- User e-mail forwarding addresses;
- User e-mail accounts;
- Local network news discussions.

Information services on the network require some underlying mechanism to transport data between user and server computers. How will the community access those information services?

- Public access terminals in libraries, schools, or other community centers;
- Modem pool: Dialup access to the network;
- ISDN, cable modems, and DSL: Higher-speed alternatives to modems that may be suitable for small groups;

- Town Ethernet utility: High-speed access from local schools, businesses, and apartment buildings. Check the costs in your area for frame relay, T1, and fiber links. Do some locations offer antenna sites suitable for point-to-point wireless links?

Providers

The goal of a community network is to deliver quality information relevant to the community; it does not have to provide all the information delivery infrastructure itself. Who will provide which services?

- *Local providers:* There may already be businesses in the area providing modem pool, e-mail, Web page, and other services. Can your community work with them to provide a direct Ethernet service utility?

- *Remote providers:* The great thing about Internet-based services is that they can be supplied from anywhere. If there is no modem pool in your community, consider one of the large commercial Internet access providers.

- *Do it yourself:* A community that has the facilities and expertise available can run its own servers, or it can start out contracting with others and switch to internal support when ready.

Who will provide operational support for the services that are offered?

- *Public access support:* Public terminals will require user education and supervision.

- *Modem pool support:* Modem pool users will require client software for their own computers and at least phone bank help getting it to work.

- *Network server support:* If the community network runs WWW, e-mail, or other servers, someone will have to monitor and maintain the server computers.

- *Network connection support:* If the network provides connections throughout the community, who will be responsible for maintaining the data lines and the communications hardware?

Where to put the equipment

If an Internet point of presence is to be established in the community, the equipment should, if possible, be placed in a location with the following characteristics:

- *Large number of in-building users:* If the point of presence is at a site such as a library, high school, or office building that has a large number of potential users, high-speed Ethernet service can be offered at very little additional cost from the outset.

- *Central location:* The point of presence should be near other sites with large numbers of potential users.

- *Secure, environmentally controlled space:* The point-of-presence network and server equipment will need about 50 sq ft of dry, reasonably clean, ventilated space behind a locked door.

Procurement and installation

IAPs supply a modem pool or connect your community point of presence to the Internet. ISPs run the Web or other network-based servers, and may also provide IAP services. Understand what your community network will be getting before any contracts are signed. In particular, seek the following characteristics.

- *Reliability:* Price is not everything! Is the provider's equipment monitored 24 hours a day? Are current customers satisfied with the quality of service? Watch out for providers that have taken on too many customers without increasing their equipment capabilities.

- *Connectivity:* What kind of connection does the IAP or ISP have to the Internet? A 56-Kbps or fractional T1 connection will bog down under heavy usage. Look for providers with at least full T1 connections (1.5 Mbps). One T1 should be able to handle a pool of 200 to 300 modems.

- *Experience:* Does the provider understand the community's needs? Try to work with providers that have experience with community networks.

- *The package:* What will your network get for the money? Will the IAP that connects the point of presence to the Internet also supply DNS and a network news feed? How many IP addresses come with the connection? Will the provider charge more if the community network in turn sells connections to local businesses? Is network hardware for the community end of the connection included?

An assortment of software, hardware, and data lines will be necessary, depending on what the community network will be offering. Here's a menu:

- *Network data lines:* A point of presence will require a leased line from the phone company between the site and the IAP. A modem pool will require a rotary of incoming phone lines for the modems. Delivery of Ethernet access across the community will necessitate connections between the point of presence and user sites: ISDN, DSL, cable modem, frame relay, T1, wireless, or direct fiber links.

- *Network hardware:* A point of presence requires at least a CSU/DSU (leased-line modem), router, and hub. If the network is running a modem pool, it will also need modems and a terminal server (preferably an access server combining the two). Ethernet access within the community will require additional equipment both at the point of presence and the user sites.

- *Server hardware:* UNIX, Macintosh, or Windows NT servers are the norm. Two medium-sized machines are far preferable to a single large one, because it will be easier to test new configurations and recover from failures. If UNIX expertise is available, use UNIX.

- *Server software:* Excellent server software can be obtained at little or no cost over the Internet, especially for UNIX. Avoid "beta-itis": If at all possible, use products and versions that are tried and true.

- *User software:* Users will need client software for their computers. A community network should avoid assembling this package itself and instead suggest commercially available or shareware packages.

Recommended resources

Note that WWW pages move from one server to another from time to time. If the URL listed here is a dead end, try looking for some portion of the item's title with a Web search tool such as AltaVista.

Internet specifics

David H. Dennis, The New Internet Provider FAQ (http://www.amazing.com/isp/).

Kevin Dowd, Getting Connected: The Internet at 56K and Up, O'Reilly and Associates, 1996.

The Internet Engineering Task Force, IETF home page (http://www.ietf.org/).

LAN technology specifics

Charles Spurgeon, Charles Spurgeon's Ethernet (IEEE 802.3) Web Site (http://wwwhost.ots.utexas.edu/ethernet/).

Charles Spurgeon, Ethernet Configuration Guidelines: A Quick Reference to the Official Ethernet (IEEE 802.3) Configuration Rules, San Jose, CA: Peer-to-Peer Communications, 1996.

Servers and server applications

Paul Albitz and Cricket Liu, DNS and BIND, O'Reilly & Associates, 1992.

John DiMarco, spectable (ftp://ftp.cdf.toronto.edu/pub/spectable).

Cricket Liu et al., Managing Internet Information Services, O'Reilly & Associates, 1994.

Evi Nemeth, Garth Snyder, Scott Seebass, and Trent R. Hein, UNIX System Administration Handbook, 2nd Ed. Englewood Cliffs, NJ: Prentice-Hall, 1995.

General references, indexes, and search tools

Compaq Computer Corporation, AltaVista: Main Page (http://www.altavista.digital.com/).

Denis Howe, FOLDOC - Computing Dictionary (http://wombat.doc.ic.ac.uk/foldoc/).

Go2Net, Inc.,MetaCrawler (http://www.metacrawler.com/).

Irving Kind, BABEL: A Glossary of Computer Related Abbreviations and Acronyms (http://www.access.digex.net/~ikind/babel.html).

Yahoo!, Yahoo! (http://www.yahoo.com/).

References

[1] Floyd County Chamber of Commerce, Floyd Community Web Page: Home, http://www.community.floyd.va.us/.

[2] Electronic Village of Abingdon, Electronic Village of Abingdon Index, http://www.eva.org/.

[3] Hinden, R. M., "IP Next Generation Overview," Communications of the ACM 39, 6, June 1996, pp. 61–71.

[4] Zakon, R. H., Hobbes' Internet Timeline, http://info.isoc.org/guest/zakon/Internet/History/HIT.html.

[5] The Internet Engineering Task Force, IETF home page, http://www.ietf.org/.

[6] MCI WorldCom, MAE Information, http://www.mfsdatanet.com/MAE/.

[7] Spurgeon, C., Ethernet Configuration Guidelines: A Quick Reference to the Official Ethernet (IEEE 802.3) Configuration Rules, San Jose, CA: Peer-to-Peer Communications, 1996.

[8] Makris, J., "Gigabit Cabling: The Copper Stopper?," *Data Communications,* March, 1998, pp 63–72, http://www.data.com/roundups/copper.html.

[9] Zimmerman, C., "Fiber Fights Back," *Data Communications,* May, 1999, pp. 89–98, http://www.data.com/issue/990507/fiber.html.

[10] Gigabit Ethernet Alliance, Gigabit Ethernet over Copper, http://www.gigabit-ethernet.org/technology/whitepapers/gige_0399/copper.html.

[11] McLarnon, B., Wireless LAN/MAN Modem Product Directory, http://hydra.carleton.ca/info/.

[12] Yahoo!, Yahoo! Business and Economy:Companies:Internet Services:Access Providers, http://www.yahoo.com/Business_and_Economy/Companies/Internet_Services/Access_Providers/.

[13] Dennis, D. H., The New Internet Provider FAQ, http://www.amazing.com/isp/.

[14] Blacksburg Electronic Village, The Village Mall, http://www.bev.net/mall/.

11

Managing Information in a Community Network

by Cortney V. Martin and Andrew M. Cohill

FROM THE START, the BEV was focused on putting people first, not technology. Developing the network infrastructure, the fiber, the digital switches, Ethernet, the modem pool, and the servers was easy compared to the task of getting citizens to use the network and the tools effectively. This chapter addresses the steps that were used to build an active, engaged online community:

1. Focus on good service.

2. Identify project champions.

3. Educate citizens.

4. Foster a rich information space.

5. Deliver the message effectively.

6. Create links between the real community and the virtual community.

Each step is critical to forming a solid foundation on which to build a thriving electronic village.

Step 1: focus on good service

The first group of BEV users was given access to the network in the spring of 1993. At that time, the BEV had no physical office space, a one-person help desk (technical support for BEV users), and a first-generation version of the BEV software. We soon learned our earliest lessons. First, it is important for people to have a physical location—an office—they can visit. Second, the demand for user support is tremendous and must be dealt with in an organized fashion.

The BEV office

The BEV office had its humble beginnings in a space in the lobby of the Virginia Tech Information Systems building. The location was convenient for the new office staff because they were close to the technical consultants and network diagnosticians. There were many disadvantages, however, to working in a space not originally designed to support an office. All the office computers were plugged into one power strip, which was connected to an extension cord, which was plugged into an outlet near the building receptionist's feet. This lifeline was frequently unplugged by an unwitting foot, thus putting an immediate halt to all office activity. However, the heavy office foot traffic proved that citizens enjoyed having a place to come and meet people and discuss the BEV.

The primary purpose of the office was to provide a central registration location. At the BEV office, a citizen could register for his or her personal identifier (PID), password, e-mail account, and network access. One form was used to register for all the services, and the prospective users signature at the bottom of the form indicated an agreement to abide by the acceptable use policy of the BEV and Virginia Tech. While the form was being processed, a prospective member filled out a questionnaire that allowed us to collect some benchmark demographic data. After processing, the user left with a PID, software, and a user guide.

The office was used informally to help and train users, answer questions, and give demonstrations. It was there that many users got their first sampling of the BEV. In October of 1994, the BEV office moved into a space provided by the Virginia Tech Museum of Natural History on Main Street. The office was separated into a public area and a closed office area, which allowed the staff to work undisturbed after office hours. The office was officially open only nine hours a week, but most citizens had little difficulty accommodating the limited hours.

The Main Street location was a real boon to the project, and the number of walk-in customers increased. The museum is a naturally inviting, people-friendly place that encouraged people to stop by and find out more about the BEV.

In 1997, the BEV group was able to consolidate all operations in a single office space. Prior to that time, the BEV public office space was all separate from "back office" technical and support staff. Having everyone together in one place made working together and providing good service much easier. Most communities, especially small ones, are likely to have only one or two paid staff assigned to the project, but office space and a Main Street presence will continue to be important. If your project expects to rely on volunteers for some help, and most community network projects do, providing decent working conditions and proper office space for volunteers will be important to training them effectively and keeping them interested in the project.

Technical support

Assistance is provided through manuals, online information, and the help desk. Even an electronic village cannot totally avoid paper manuals, at least not yet. On average-size monitors, it may be difficult to arrange the necessary windows, icons, and dialog boxes to enable users to follow online instructions. For BEV customers, our design targets inexperienced users who may be more comfortable with traditional media. We distribute a paper manual that is supplemented with online material. Supplementary diagnostic materials and FAQs are available on the network. We have found that ISPs often provide only a bare minimum of information and often assume a level of technical experience that most users do not have. An important role for the community network is to provide additional information and training opportunities to supplement private sector efforts.

As we were designing documentation for multiple platforms, we focused on streamlining and reusing content. When we began offering packages for the three major operating systems (DOS, Windows, and Macintosh), we distributed platform-specific installation guides, that covered all of the steps necessary to install and configure the software. An accompanying user guide covered content common to all platforms, such as material on the history of the Internet, Netiquette, and using the clients.

Over time, the client software evolved to the point where the interfaces were almost identical across platforms. That enabled us to move more information out of the installation guide and into the user guide. The value of that model is that the installation guide can be reduced to just a few pages, which are inexpensive to reproduce. The lengthier user guide can be distributed through the local copy center, the public library, and the network.

We saw no reason to try to compete with the many books published on navigating and using the Internet. Our user guide concentrates specifically on local issues, such as acceptable use policies, posting information, joining the local listservs, and finding local e-mail addresses. The user guide also addresses the specifics of the operation of our servers and our network.

The first line of support for most users is the help desk of the local ISP (modem pool provider), but many users will still need additional help beyond that available from the telephone support these companies usually provide. There was a high demand for house calls to solve problems. Our limited staff could not provide enough support, so we created a consultants Web page. Both volunteer and for-pay consultants could offer their services, and the page was used frequently by help-seekers and consultants. Interestingly, we found that people were reluctant to pay the consultants upfront, possibly because there are no clear indications of their qualifications, no certification, no assurances. The more successful consultants had a policy by which customers paid only if their problem was solved.

The documentation was refined through informal usability testing. Although we did not have the time or the budget for formal testing, manuals went through many iterations in an attempt to eliminate confusing or misleading information. Many mistakes were caught that saved hundreds of users from calling the help desk.

The office, the software, and the technical assistance were all important factors in getting the citizens of Blacksburg to the brink of the online community. However, we needed quality information to lure them in and keep them interested. Early in the project, we needed advocates who wanted to use this new publishing medium.

Step 2: identify project champions

Our project champions offered an abundance of energy, creativity, and effort to the project, especially in the early stages. They were important for giving the project momentum and for setting a good example.

An electronic village is not just a connection to the Internet. It is a group of geographically colocated individuals, interacting electronically with each other, with local content, and with the worldwide resources of the Internet. Where does that local content come from? Well, without a large staff whose only purpose would be to gather and post the information, we had to rely largely on volunteers, and many of our early volunteer contributors were project champions.

Online information should mirror what is found in the true community. In Blacksburg, a community is made up of groups like the following:

- Businesses, the Chamber of Commerce, professionals, and self-employed individuals;

- Health organizations, doctors, and therapy groups;

- A library and museums;

- A tourism bureau, a historical society, and steering committees;

- School administrators, teachers, students, parents, and the PTA;

- Government agencies, elected officials, a city council, a transit authority, and social agencies;

- Clubs, organizations, and civic groups;

- Churches, synagogues, and religious organizations;

- An arts council, galleries, and local artists and entertainers;

- Local newspapers and magazines;

- Senior citizens, schoolchildren, and families;

- Individuals.

A well-balanced community has representatives from each of those categories. Some community networks may unknowingly exclude one or more of the groups; businesses are the most frequently left out. Unless everyone has a place on it, the community network will become fragmented and of much less value to the users.

Each category of community members should be targeted as a potential source of information and resources. It just takes a few dedicated individuals and groups putting up well-designed content to attract more contributors to the electronic village. The BEV has had countless champions, far more than can be mentioned here. For brevity, we will highlight only a few and mention their contributions that inspired others:

- The BEV Seniors;

- The Town of Blacksburg;

- The Volunteer Action Center (VAC);

- The YMCA Open University.

The BEV Seniors were among the earliest and most enthusiastic users of the BEV, and surprisingly, they were among the most willing to try and do new things. Their BEV Seniors Web site http://www.bev.net/seniors/ is consistently one of the most vibrant and active sites in town. As the number of ISPs in town grew, it became increasingly difficult for consumers to determine what companies were offering services, what the rates were, and how to contact them. The BEV Seniors began publishing a summary page listing all that information, and it is still heavily used today as more connectivity options like cable modems and xDSL services become available. The BEV group keeps paper copies of the page in the BEV office and hands them out to residents who stop by the office trying to figure out how to get connected.

Because it is one of the partners in the BEV, the Town of Blacksburg's online information (http://www.blacksburg.va.us/) contribution was critical to the success of the project. Citizens watched their progress carefully as a barometer of project success. Town officials and staff had the ability to make a great deal of information available to the public, information that people sought. How do I reserve a picnic shelter? What does a dog license cost? Will trash be collected on New Years Day?

The town designated one information provider in each department who was responsible for uploading and maintaining departmental information (Figure 10.4). It had an organized three-part plan for getting information on the Web and Gopher. Phase I included making textual, noninteractive information available. Phase II involved adding Web-based interactive features: forms for picnic shelter reservations, police vacation check applications, and so on. Phase III includes network-based financial transactions: paying the water bill and buying a town car sticker. The last phase has been delayed by the lack of a good, secure transaction system. A secure Web server encrypts all the information (like credit card numbers) that travels across the Internet between a user entering the information and the merchant or agency receiving the information. The town soon began publishing departmental e-mail addresses in the town newsletter and gave URLs for related information. It is the interweaving of traditional media and electronic media that encourages people to integrate the electronic village into their lives.

In 1998, the VAC (Voluntary Action Center) came to the BEV with a problem. In Blacksburg, the VAC maintains a master list of about 250 social service and volunteer agencies to assist local social workers, church groups, and other volunteer agencies in matching people in need with the right organization. The list was expensive to print, in high demand (they circulated several thousand copies once a year), and difficult to update.

Using our database to Web publishing tools, we were able to create an online database of the VAC information (http://civic.bev.net/vac/) that could be accessed and searched from any Web browser, freeing the VAC from the high cost of printing and distributing paper. We also created a PDF version (a downloadable print file) that allowed anyone who needed a paper copy to download it from the Web and print it as needed.

Finally, we created a Palm Pilot version so that Palm Pilot users could download a file, export it to their Palm organizer, and have a personal electronic version even if they did not have convenient Internet access.

This joint publishing effort by the VAC and the BEV eliminated countless hours of hand maintenance of the data, enabled them to provide more information to more people, and added important new features like full-text searching and multiple indexes so that needed services could be found quickly and easily. This is an excellent example of how the community network group plays a synergistic role in the community by providing technology expertise to civic groups that could not afford to this kind of advanced publishing any other way.

The YMCA in Blacksburg runs a very popular Open University, which offers free and low-fee short courses and seminars on a wide variety of topics, including health, arts and crafts, technology, dance, foreign languages, and many others (http://www.vtymca.org/). The YMCA provides a full list of all their classes online and even offers online registration with the assistance of a local ISP that provides a secure server for that purpose. Formerly, registering for a class required making a special trip to the YMCA office to sign up, but busy mothers, working fathers, and anyone who wants to save a little time can register online.

The BEV is a success today because of the champions mentioned here and countless others, big and small. Click on any page making up the BEV Web site, and you will see a valuable contributor, whether it is the senior citizens' page editor, a laundromat owner advertising a special, a local physicians' medical database, or the New River Arts Council's listing of local bands. We never know who the next champions will be. We could find them in public offices, in the factory, in the schools, or among the senior citizens. Whoever they are, we work to keep their motivation high and to provide them with the tools and training they need. Their achievements go a long way toward ensuring success of the entire project.

Step 3: educate, educate, educate

One measure of success for the BEV is the number of citizens who are online. As with the telephone, it is important for people to be able to interact with other people who are important to them, whether that is their children's teacher, businesses, or friends and neighbors. The more people who use the network, the more valuable it will be to each of them. How did we attract people to our community network? A multifaceted community-wide awareness campaign was used to expose citizens to the technology and to show them how it could be useful in their lives. We employed a concurrent top-down and bottom-up approach to community education.

The top-down approach involved getting information about the BEV out to a large audience. Television news, radio programs, and newspaper articles enabled us to reach the largest number of people, but those media lacked the depth and interactivity that was needed to put people at ease with the technology. Early descriptions of the BEV may have aided a perception of the project as something out of a science-fiction movie, or at least something that required special knowledge and training.

The next tier down targeted town forums and speaking engagements at clubs like the Kiwanis or the Newcomers Club. Citizens did not get individual hands-on experience with the network, but they did see live demonstrations and were able to interact with the presenters. Rather than offer dry descriptions of the tools, we found it useful to present the technology in terms of real-life situations, for example, a child struggling with her homework. Using the network, the presenter would show how the technology could enable a parent to address the problem in a new and efficient way. The parent might go to the child's school's Web pages to see what has been posted about homework: tips, alternatives, references, and so on—or the parent could send an e-mail note to the teacher, explaining the child's frustration and asking for help or requesting a meeting.

The case study demonstration made a more lasting impression than a straight showing of the Web, e-mail, and Usenet. We always tailored demonstrations to the audience and did not hesitate to use people's competitive instincts for leverage. For instance, we would show the Kiwanis Club what the local Ruritans had done. Then, we would use a search engine to generate a list of all the Kiwanis Clubs online throughout the world. It is amazing how much enthusiasm can be generated in a group that feels left behind! The top-down process really just exposed the community to the BEV. To actually get citizens to commit time and resources to the project, we had to work from the bottom up.

Our early efforts at a bottom-up approach to educating the public was a grass roots effort. The BEV staff frequently made house calls to help citizens who were having trouble with software or just could not figure out what to do with it. Doing so was labor-intensive, but in some cases, it was the best solution. The voice of a disgruntled customer reaches far and wide and can be damaging to the project and to citizen morale.

It became clear that one-on-one training was not a good long-term solution because it is so labor-intensive. A more efficient solution to hands-on training was offered by the staff at the Blacksburg branch of the MFRL. The library has six public access computers with direct (Ethernet) connections to the BEV and the Internet. The library offered training for e-mail and the Web. The classes were limited to six attendees, one per computer, which allowed the instructor to offer individual attention. Students were assisted in finding information useful to them, whether it was gardening, genealogy, or geology. The library trained more than 1,000 people over an 18-month period, six at a time.

We identified a number of groups in the community that could benefit from targeted training: town employees, local businesspeople, and school teachers. The town employees were trained primarily by the town BEV liaison in small group sessions. To encourage local businesses to get online, we offered a number of three-hour seminars, which we advertised through a Chamber of Commerce mailing. The seminars, which were completely filled and even had waiting lists, were held in a computer classroom with each student computer having a direct Internet connection. The seminar consisted of introductory material and a demonstration showing the way businesses could adapt their services to the Internet. In the second half, participants used the network while a team of BEV instructors met individually with each person to brainstorm on how his or her business could benefit from the BEV. Many ideas were generated in those seminars, and many of the ideas were implemented. Visit the Village Mall (http://www.bev.net/mall/) to see examples of what Blacksburg local businesses have accomplished on the Web.

While Internet connections to each of the county's 19 public schools were being established, a plan was under way to help teachers learn to use the new Internet tools in the classroom. The county identified a core group of teachers to be trained who would then return to schools and share with other teachers what they had learned. Recently, a state-of-the-art computer classroom was installed in one of the rural schools to be used to train students, teachers, and parents. See Chapter 7 for more information about computers in the classroom.

In the past, the BEV staff was the major source of training programs for the community. With limited staff, we could never meet demand, and we encouraged other individuals to offer instructions. Programs began to be offered through the YMCA's Open University, through the Town of Blacksburg's Department of Parks and Recreation, and through Virginia Tech's New Media Center. The range of classes has grown and now includes basic Internet classes, modules on the Web, e-mail, and introductory and advanced HTML classes. Six years after the project started, demand for training has not decreased. The topics we teach are now different—there is little demand in the community now for classes on e-mail, but there is high demand for classes on how to use Internet search engines.

Our computer-literate community has a number of highly qualified HTML volunteers who are eager to create pages for any organization. Our backup plan, if we need to recruit more Web help, is to offer free HTML classes in exchange for a certain amount of authoring service. Part of an in-class exercise would be to have students create Web pages for assigned organizations in town. The general information provided would be club purpose, meeting schedule, officers, upcoming activities, and perhaps a photo. Participants gain experience in Web page authoring, and organizations get a Web presence.

Although we did not have the resources to incorporate them into our training plan, videos could be a good instructional supplement in a community. They can reach a large number of people, especially if distributed widely. They can be put in the public library, shown on the community cable access channel, shown in classrooms, and sold. Videos are time-consuming to produce, and it is not easy to make a clear, understandable program that anticipates and answers more questions than it generates. However, if successfully created, a video program could really ease a community's instructional deficit. A community with a public access TV station might be able to partner with the community network to produce instructional videos.

With the introduction of Quicktime 4 by Apple Computer in 1999, video streaming (sending video over the Internet) is now both affordable and technically in reach for many organizations. Apple has provided open-source (i.e., free) server software for streaming video and compressing video for downloads, making it likely it will be adopted widely. Providing video on a Web site is still limited largely to users with adequate bandwidth, but short video clips (under 30 seconds) can be downloaded in a reasonable time by users with 56-Kbps modems.

It is now possible to create short video clips on a variety of technical and content development topics that people can download and watch at their convenience. One must keep in mind that not everyone will wish to do this, and it may still be necessary to create and provide videotapes.

We are fortunate to live in a town with several computer classrooms with direct Ethernet connections, but that is not a requirement of successful instructional programming. A computer with a 56-Kbps modem connection and an LCD projector is a worthwhile investment for any project. The equipment can be used for instruction or demonstrations in town or on the road. If possible, a community network should invest in at least a small cluster of Internet computers that can be used for public access and/or training. As mentioned earlier, many people can be trained on only six library computers.

Step 4: foster a rich information space

We readied the tools, identified champions, contacted content suppliers, and began training the townspeople. So what would the citizens of the BEV be able to do when they were online? We worked to create a rich environment in which users can interact with information and each other. The BEV offers the most common Internet tools:

- E-mail, including local and global listservs;

- A Usenet newsreader;

- A Web browser;

- FTP and Telnet.

FTP and Telnet are used infrequently and will not be discussed here. This section will address the ways we facilitated the use and effectiveness of the other tools listed. Internet activity somewhat mirrors communication in the physical world. We most frequently engage in one-to-one or one-to-small-group discussions, whether in person or over the phone; e-mail is the corresponding medium in the electronic world.

E-mail

E-mail has the obvious advantage of being asynchronous, that is, the participants do not have to be engaged at the same time. One user can send a note out in the morning, and the respondent can reply in the afternoon. We have found that users enjoy e-mail because it allows them to regain some control over how they spend their time. They are not slaves to a ringing phone or doorbell. Some people, however, fear that e-mail will replace face-to-face communication. With the BEV, we have found quite the opposite: E-mail seems to complement face-to-face meetings. For example, our BEV-Seniors senior citizens group uses e-mail to democratically decide on meeting and social times—the online conversations seem to drive the need to get together.

E-mail also facilitates the process of getting to know new people. One is less likely to prejudge based on appearance, age, gender, weight, height, and so on. There are countless stories of citizens finding common bonds with others using e-mail, with true friendships ensuing. A BEV staff member related the story of meeting one of her own neighbors as they exchanged e-mail about her Web site. The neighbor recognized her name from her e-mail and remembered it from her postal mailbox on the street. A connection was made that probably would not have happened in a meeting on the street. Members of the BEV Seniors group offer another example. Few of them knew each other prior to meeting via e-mail and the listserv. Many friendships have resulted from their electronic communication.

The e-mail experience can be enhanced by offering training and easy-to-use e-mail directories. We mistakenly believed that e-mail was simple enough for citizens to learn to use, so we did not focus our early education efforts on it. Senior citizens were the first group to lobby for e-mail training. They wanted to fully understand mailboxes, nicknames, and signatures. In our first training and question-and-answer period, we realized that many people did not fully understand the client-server model of computing, which is important in understanding where the mail is and what happens when you check mail on more than one computer. The e-mail training is perceived by trainees as worthwhile, and we anticipate that it should reduce the number of related help-desk questions.

An online directory is important, and it should be searchable by first name, last name, and e-mail address. Our directory is available through the e-mail client or the Web browser. The Web directory is nice because it uses a simple Web form and includes basic instructions on how to search using wildcards when exact spellings are unknown. Citizens can elect to keep their information private, but only a minority actually choose to do so. We offer links to other commonly queried directories of nearby communities, schools, and agencies.

Ideally, directory services should mimic how people search each other out in real life, for example, seeing a name on a mailbox or a phone number on a team roster. It is not unusual for a person to ask questions like, "I wonder who on my tennis team has e-mail?" The BEV introduced a sophisticated new community directory system in 1999 that is being used widely by the community.

Listservs

Listservs, or electronic mailing lists, manage and maintain central lists of e-mail subscribers to lists devoted to particular topics. There are global, worldwide lists covering everything from tomato gardening to fat-free cooking and local lists limited to particular network domains. Our users have access to the global lists as well as a number of local lists.

The first local list was BEV news, a list intended for discussing BEV activities like research seminars or Internet training schedules. For a time, that was the only BEV list available, so the content was never well-focused, and it was used as a catchall for any topic that was brought up. The list is unmoderated; that is, no one screens the submissions. Topics range from highly technical discussions, to arguments about auto repair facilities, to general chat. It was on this list that many people learned about flaming. From time to time, we would occasionally post the rules of the list, which were derived from our acceptable use policy. We let the list govern itself, and it eventually stabilized somewhat. The list lacked purpose, lacked strong leaders, and lacked a format. We learned we must offer lists that are narrower in focus.

One of the next (and narrower) lists to develop was the BEV Seniors list, intended for those who are approximately over the age of 55 and live in Blacksburg and vicinity. We created the list about the same time that we carved out a Seniors page on the Web. The list started off slowly, with infrequent messages saying something like, "Is anybody out there?" Fortunately, several senior champions emerged. Online introductions began, and everybody got to know each other. The list membership was monitored, and as new members joined, they were invited by the informal welcoming committee to say a few words. Fortunately, one of the most active seniors volunteered weekly in the BEV office. There, he was able to explain the listserv one-on-one to prospective members and encourage them to join. Listservs are not inherently easy to understand, and they can be intimidating. Though the BEV Seniors membership was growing, only a small number of members were contributing to the discussions. At a subsequent live meeting, we learned that many of the seniors were not sure what was appropriate subject matter, so they just did not post anything. After they had reassured each other that it was okay to post notes about how one's tomato plants are faring, about one's granddaughter getting into the cat food, or about what a beautiful day it is, traffic on the list increased significantly. Some of the women in the group expressed concern that only the men were contributing. That has changed somewhat, and there are now more women on the list than men, although a disproportionate number of men still post messages.

The BEV Seniors list became so successful that the members expressed concern about protecting the integrity of the list. Several nonlocal and nonsenior citizens joined the list. The configuration of the list was changed so that a prospective member was screened before being accepted to confirm that he or she was indeed a local senior. After receiving the confirmation, the list owners added the person to the list.

The listserv members have had a few problems, most of which were simple misunderstandings of tone and intent. One member posted some interesting facts about the makeup of the Earth's population. A reply to the note violated an oft-written Internet rule: Do not reply in all uppercase letters, which is usually interpreted as electronic yelling. Predictably, the all-caps reply was misinterpreted, and some feelings were hurt.

As in an actual conversation, the listserv fell silent for a time until some uncomfortable members posted innocuous friendly notes to initiate new discussions. Listserv participation requires a high level of care, higher than many realize.

The seniors have been able to accomplish a great deal using e-mail and the listserv. They schedule meetings, participate in surveys, and help members with technical problems. They participate in an electronic pen pal program with an elementary school. They are writing about Blacksburg nostalgia, possibly to be published on a CD-ROM. They design curricula for their computers for codgers classes. One member who winters in the West with his son stays in touch with friends via the listserv.

The BEV Seniors group has indeed created a strong virtual community among themselves. This virtual community has not replaced the physical community but rather has enhanced it. Many of the BEV seniors did not know each other prior to joining the list. At the monthly meeting, one gets the sense that these people have known each other for years.

When setting up a listserv, a key decision is whether to make the list moderated or unmoderated. With a moderated list, only the list owner is allowed to post messages. Everyone must send messages to the moderator first, who has the option to reject the message, edit the message and post it, or post it as is. Moderated lists tend to have fewer messages of higher quality. For civic groups in Blacksburg, we recommend using a moderated list for announcements, newsletters, and important messages from group leaders.

Unmoderated lists allow anyone subscribed to the list to post. Unmoderated lists tend to have more messages of lower quality than moderated lists, and many people do not wish to be subscribed to unmoderated lists because this kind of list tends to generate a lot of e-mail. If civic groups want to use an unmoderated list to encourage chat among group members, it should be a second list, in addition to the moderated list. Many group members will not stay on an unmoderated list, making it useless for sending out important group news like meeting announcement and minutes.

The following ingredients seem to be key for a successful community listserv:

- A narrow focus;

- A common thread among members;

- Vocal champions to stimulate participation and solve problems;

- Member control of the list membership;

- Local members only.

Dozens of community groups in Blacksburg now use mailing lists to help manage their activities.

Newsletter

In October of 1994, the BEV began distributing a monthly electronic newsletter using e-mail. We have a mechanism in place for sending e-mail to all the BEV members who registered at our office. Almost no one has objected to the once-a-month electronic mailing; most people enjoy receiving the newsletter. The main purpose of the newsletter is to distribute information about BEV services, but it also reminds people that they are part of an exciting project. Content varies from month to month, but includes things like the following:

- Office registration statistics and the BEV volunteer of the month;

- Procedural instructions (e.g., canceling service or receiving technical help);

- Statistics on top Web pages and top personal home pages;

- New services (e.g., listservs, new pages, and newsgroups);

- Schedule of community Internet training classes;

- Local Web page hunt;

- Reports from the schools, town, researchers, and senior citizens;

- Letters from the BEV project director.

The newsletter is an ideal vehicle for distributing information and generating enthusiasm, and it acts as an informal project history. For visitors or users who may not be on our e-mailing list, the newsletter receives a temporary link on the home page when it is e-mailed. Afterward, the newsletter is archived on the Web. Figure 10.7 is a sample portion of the BEV newsletter.

Usenet news

Almost since the inception of the project, the BEV has offered Usenet newsgroups. In fact, we had about 10 newsgroups before we were even distributing newsreader software. Traffic remained somewhat low even after we began including news software in the BEV distribution. We attributed the lack of activity to software problems, unclear group purposes, and too many groups.

The newsreader feature of Web browsers is not yet adequate—it is hard to use and manage. Users can elect to use the Web newsreader or can start up the separate news client. The effort involved is often enough to deter many users.

We developed the news hierarchy before we understood what it was that people would want to talk about. There were too many groups that were too general. Many people told us that they were uncertain about where to post their messages, so they just did not post them. In response to that, we reorganized our newsgroup hierarchy.

Groups can always be added to the hierarchy, but it is much more difficult to remove groups, which we unfortunately have had to do. Much of the group content overlapped so much that there was a great deal of confusion about what went where. We recommend that community networks start with a small number of well-defined groups that have as little overlap as possible. We settled on seven Blacksburg groups: bburg.announce, bburg.general, bburg.business, bburg.config, bburg.environment, bburg.forsale, and bburg.k12. The two BEV groups are bburg.bev.announce and bburg.bev.general.

Interestingly, several very specialized newsgroups spun off early on. They included a local auto repair group (bburg.auto.repair), a beer brewing group (bburg.hobbies.brewing), and a sports group (bburg.sports.general). Traffic on those has tailed off as well, perhaps because the groups are too narrow in focus.

We may have been better off starting with just two groups, such as bburg.general and bburg.bev.general. With two well-defined groups, it would be easier to decide which newsgroup to use for what. One would be for BEV-specific posts, one for general town talk. It would reduce users' fear that they are not posting in the proper place. As confidence builds, traffic should increase. That traffic would encourage even more posting as people perceived a higher probability that they would receive a response.

Another technique we used to increase activity on our newsgroups was to gateway our BBURG-L listserve to the bburg.general newsgroup. Users can choose how to interact in this group designed to address issues of the town. Anything posted to the list is cross-posted on the newsgroup. Anything posted to the newsgroup is sent to the list subscribers.

If local Usenet groups are created for use by children, it may be necessary to have volunteers and/or paid staff monitor the groups to ensure that discussions remain on age-appropriate topics, to ensure that children observe proper Netiquette, and to discourage name-calling and bullying behavior.

Usenet is the virtual town hall, permitting threaded conversations on any topic. We have seen some heated discussions about a new highway, the new Super Wal-Mart store, and local elections. We are working toward a more thorough integration of government officials in the virtual town hall. We envision a day when the town council agenda will be posted and threads started for each topic. Council members could oversee discussions, correct misconceptions, refine their questions, and inform the public. In 1998, the BEV began using an online conference system that offers much better usability than chat for civic discussions, and the BEV began experimenting with online conferences about local civic issues in 1999.

The Web

The golden rule of the BEV Web is to maintain a tight focus on community information. With the availability of large-scale Web indexes and Internet-wide search engines, it makes little sense to try to compete in those areas. The greatest success came from concentrating on our niche: information for, about, and by our citizens. That philosophy kept the project team focused, which is critical with limited resources and personnel. Several other philosophies guided the creation of our online information:

- Focusing on local content;

- Striving for quality over quantity;

- Using a carefully designed Web page template;

- Encouraging contributions;

- Designing for the user with minimal equipment and knowledge.

There are many good books and online resources that provide guidance on good Web design. A working knowledge of HTML (which many people have) does not automatically qualify an individual as a competent Web page designer. Many communities consistently make the same easily avoidable mistakes with their page design, including the following:

- Too many pictures, too little content;

- Very large pictures that take a long time to download;

- Flashing pictures and animations (should be avoided at all costs);

- Lack of coherent graphic design;

- Pages that consist of nothing but links (many of which are broken);

- Too many pages marked "under construction."

The best place to start to understand more about Web design is Jakob Nielsen's Web site (http://www.useit.com/), which has much useful information and links to many more. Careful design of the community site is a critical issue that should not be left to someone's nephew, the local high school computer class, or a volunteer interested in on the job training. A community would never let the new town hall be designed by any of those groups, and the community Web site should receive careful attention. Design professionals may be willing to help out if asked, or it may be worth hiring someone with a proven track record (ask to see several sites he or she has designed) to provide the community with a set of templates.

Locally generated information is worth more than its face value because it is backed by the integrity of fellow local citizens. A good example is the BEV's online medical database, which is small by most standards, with just over 100 text descriptions of common medical conditions and recommended treatments. More comprehensive medical databases certainly can be found elsewhere on the Internet. The real value of the BEV medical database is that it was created and is maintained by a local physician, a physician who is known and trusted throughout the community. The database is strictly informational; it includes a disclosure that advises users to consult their physicians before acting on information in the database.

From the beginning, we stressed quality information over quantity. It is easy to publish volumes of information on the Web, but consumers are quite discriminating. If a page looks poorly organized or unappealing or takes too long to load, users will stop the download and go on to another page, and they may never return. We started small even when it meant having only a few community organizations on a page. We found that a small number of well-done pages set a higher standard and better demonstrated the potential of the medium. It amazed us to see how quickly the content accrued.

It is important that citizens have one place on the Web that fairly represents the entire community without any particular bias or business motive. The site could be run by a nonprofit group, the town government, the chamber of commerce, or, in our case, a university (see Figure 10.8). This unification is the quality of the BEV Web that is praised most often by local users and visitors.

Most civic groups maintain their sites on the BEV server, but some groups use other servers in town and rely on the BEV Web site to provide the primary link to their information. The other Webmasters alert the BEV staff to new links to ensure the BEV listings are complete. Members know that the Village Mall and community organizations' listings are the definitive page.

The BEV is a two-way communication network, not a one-way information delivery system, like a TV set or a radio. With the BEV, every member is encouraged to be an information provider as well as an information consumer. The BEV Webmaster's job is not to create the content but to organize and present the resources in a usable fashion. It is important to have tools that allow citizens with a wide range of computer skills to contribute information. The primary purpose of the community

Web site is to provide a structure for all information in the community, no matter where the information is hosted.

We design our Web pages for the user with the minimal equipment configuration and knowledge. We recognize that many users rely on older computers and slower modems. We strive to keep pages small and with few graphics. We do not necessarily incorporate the newest HTML tags or the latest features (like Java), which may be troublesome to older Web browsers. If we do choose to add the latest features, we also offer alternative lower tech pathways through the BEV Web. For example, the BEV Web site uses no Java and almost no Javascript because both languages still make browsers and/or computers crash too frequently.

Step 5: deliver the message effectively

Web design is a complex process of creating a high-level structure, mapping out navigation routes, laying out a template for pages, and creating individual pages. Our organization techniques are derived from human factors engineering principles. We do not always have a great deal of time to devote to our Web site, so we sacrifice flashy graphics and image maps for organization, consistency, and simplicity.

We had a general structure for the Web in mind from the start, even though content was slow to develop. We used placeholders for the essential pages of our community Web site: businesses, education, clubs and events, government, people, and project information. We tailored our Web to our community. For instance, we have a page on area museums that features the Virginia Tech Museum of Natural History.

A template is the basis of all the major BEV pages. It includes a unifying logo at the top and a common tool bar at the side. Each page also has the URL, a revision date, and the page editor's e-mail address.

We try to help visitors determine what is new on the site. Return visits are more fruitful because users can see immediately what has changed. We draw attention to new entries with a new icon. It is also a nice way to focus attention on new information providers. The home page is designed to use about two-thirds of the space as a public bulletin board for timely notices of meetings, events, and happenings in the community.

On lengthy pages, we put the most important information on the first screen. We strive to give people enough information so they can immediately determine if they are in the right place without having to scroll through the page. (A page with a large image at the top followed by a small amount of information several screens down may be visually interesting, but it can be frustrating to a user looking for specific information.)

The BEV Web does not serve pages composed entirely of links. Pages of links with no real information can be boring and daunting. Such a dry list can easily be transformed into an interesting page by pruning the list and providing some narrative about the page and perhaps a brief description of some of the links. Search time can be reduced by categorization of the links to break them up. Long lists of links also get stale, and so require a lot of work doing regular link-checking to keep them up-to-date.

On long pages, it is helpful to provide a quick index. For instance, the Blacksburg YMCA publishes its course offerings for the Open University on the Web: cooking classes, dance classes, arts and crafts classes, and so on. At the top of the page, a quick index reads:

[Cooking] [Dance] [Arts and Crafts]

Keep the quick index small. Running the key terms across a line is just fine. It is also helpful to duplicate the quick index at the bottom of the file or, in some cases, in the middle of the file.

We make it easy to move forward and backward among related pages. Some documents are best published in sequential chapters or pages. Our Vision Statement is on multiple pages but rather than require users to return to a table of contents each time they want to get to the next page, we include links to next section, previous section, and table of contents.

Similarly formatted links within a single page should lead to similar units of information. The BEV Community page has links to community organizations and, sports and religious groups. Each link leads to a comparable unit of information, though they may be of slightly different sizes. The Community page also contains links to the large New River Arts Council Arts and Entertainment site and an elaborate Senior Citizens section. Those pages contain many more links and are fundamentally different from a unit of information associated with a club. To emphasize the differences, the links are indicated by prominently labeled graphic buttons.

With lengthy listings, we make alternative indexes available. The Village Mall has links to a categorical index and an alphabetical index. The BEV local search tool is also available from all the major pages. Links can also be organized by access frequency, date of entry, or size.

Of course, the most essential part of the community Web is the lower level information about groups like the Boy Scouts, the local grocery store, and the neighborhood church. We have developed a number of methods to help them get their information out to the community.

Helping contributors get their information online

Information providers have a variety of methods for getting information on the Web, depending on their expertise and needs: e-mailing entries to the Webmaster, completing a Web-based form for automatic page generation, or FTPing to a filebox.

E-mailing merely requires a contributor to send a preformatted text entry in the body of a message. The content is manually placed into the Web structure. This procedure is too labor-intensive if there are a significant number of entries. To reduce the overhead in getting information online, we are moving toward pages that are generated automatically by the contributors using a Web-based form to enter information. The structure of community information lends itself to automation. A club generally will want to post its mission statement, list of officers, meeting times, and a page of related links. A business will describe its product or service and include a phone number, e-mail address, hours, and address. Individuals want to list their hobbies, favorite links, and maybe some personal stories. We secure entries with the users' BEV PID and password; only the user can edit the information he or she has entered.

In early 1999 the BEV began using a Macintosh-based database to run the Web publishing system for major portions of the Web site. By this time, the hundreds of business and community groups began to overwhelm our ability to maintain the links and pages by hand. Business and community group links are now stored in a database, and each entry is maintained by the business or civic group, rather than by the BEV group. Link owners log into a Web page form using a private userid and password and can change and edit the title of the link, the link (URL) itself, and a brief explanation about the group or business. Using this automated system reduces the amount of time needed to maintain the Web site and allows users to update the links and post them online immediately (instead of sending e-mail and waiting for the BEV group to do it). An added benefit is the ability to provide very fast full-text searches of links and to provide multiple indexes and link sets. This system is available under license from the BEV.

The new system also provides a Community Directory, where residents can register their e-mail addresses, the URL of their personal Web pages if they have one, and other information like a phone number if they choose to make that public. Each registered user can also create a personal Web page directly on the BEV server.

Individuals can use the BEV home page Web form to create a simple or advanced page (Figure 10.9). To do so, they select *Create a home page* on the BEV Villagers page (http://www.bev.net/people/). Users cannot upload and post their own graphics because of the storage space involved and because it would be difficult to manage. Two ways to enable graphic support include giving everyone an individual FTP account or giving access to a public FTP site. The former method is labor-intensive for the system administrator; the latter way has the drawback that users can delete or modify other people's graphics. Home page authors can use HTML to point to graphics elsewhere on the Web. Users do not need any special skills to create a page using the Web forms. These pages may not be used for commercial purposes.

If individuals do not want to use the home page creator forms, they can opt to use one of several servers elsewhere on the Internet offering space for home pages and sometimes graphics. A Web search tool will find home page services that are available. Also, individuals can contract with a Web server business. Many ISPs bundle Web server space into their Internet service package. Otherwise, it can be purchased separately.

The BEV offers civic groups a package called Community Connections. A group registering for Community Connections receives a full service Web site with FTP access, two permanent e-mail addresses, and a mailing list for use by the group. The Web site has 10 megabytes of storage space and is indexed by the BEV search engine. An annual fee of $20 is charged for this service.

In 1999 the BEV began maintaining an online calendar of its own for dated material. Citizens who want to post dated events may also do one or more of the following:

- E-mail the event notice to the New River Current to have it printed in the newspaper and posted on its online calendars, available on the community page;

- Post it to local Usenet newsgroups like bburg.announce or bburg.civic;

- Post it to a listserv like bburg-l or bev-news;

- Announce it on a community groups Web page;

- Electronic classified ads (for sale or item wanted) are not posted to the Web as such. They should be posted to a Usenet newsgroup like bburg.forsale. A major event affecting the whole community can be sent to the Webmaker for possible inclusion in the message of the day section of the BEV home page.

Businesses may get a free, but very brief text listing for their registered Montgomery County, Virginia, business by using the BEV automated Village Mall registry. If businesses want Web pages or enhanced services, they must contract with a commercial Web server or ISP to create a Web site. The BEV Village Mall provides a business category for companies that provide Web hosting, since this is a common inquiry in the BEV office.

Regardless of where an organization's Web page is located, the page can be linked up to the BEV Web on the Community Page, Village Mall, or other related page. All that is required is the URL and organization name; the business or organization must serve primarily the population of the New River Valley to be listed on the Community Page.

Note that information providers need not have network access from their home or office. All members of the community are eligible for free e-mail accounts and can use the public access computers in the Blacksburg library. Those computers have all the tools necessary for users to access e-mail, Web forms, or fileboxes.

Choosing the right vehicle

There are many kinds of information to share in a community network, and there are many ways to serve that information: e-mail, the Web, listservs, and Usenet newsgroups. We try to pick and choose the best vehicle(s) for each piece of information.

The information is evaluated for criticality, frequency of change, and intended audience. Those characteristics guide in the choice of the proper vehicle for delivery.

The delivery vehicles differ by their frequency of use and their intrusiveness. In our community, e-mail is used most frequently, while Usenet newsgroups are used the least. E-mail is more intrusive than a Web posting. Users can choose not to access the Web or not to read a particular entry, but users cannot easily choose to not accept mail. The e-mail program downloads all the mail and cannot filter it from the server side. It is easily deleted, but it is nevertheless an intrusion.

Information varies in importance from critical (e.g., anticipated network outage) to noncritical (e.g., recreation center hours). Table 11.1 ranks several delivery mechanisms in terms of how well they serve critical information.

In the event of a highly critical event, a mechanism is in place to e-mail all BEV members. Few events warrant this highly intrusive action. Most important events receive sufficient attention by being included in the bulletin board area of the BEV home page. The next avenue for

Table 11.1
Delivery Mechanisms

Urgency	Delivery Mechanisms
More critical	Emergency e-mail to all BEV members
.	Posting to the bulletin board area of the BEV home page
.	E-mailed in monthly newsletter for all BEV members
.	Announced on local listservs
.	Posted to local Usenet newsgroups
Less critical	Standard entry on the Web

communication is the monthly electronic newsletter that is e-mailed to all BEV members who have registered through the office. We feel that one BEV mailing per month is not unreasonable.

Notices can be posted to the local listservs that serve a subset of the BEV membership. Again, with e-mail, it is likely the message will at least receive a glance. With the local Usenet newsgroups and the lower levels of the BEV Web, the posting may or may not be seen.

Another characteristic of community information is how frequently it changes. Some information almost never changes (i.e., a historical record of the town and the mission statement of a local club). Other information changes almost daily or weekly—for example, a calendar of events, new video releases, or minutes of the town council meeting. Information such as a description of a business or bus schedules might change only yearly. Frequency of change is important in determining who should maintain the information (information manager versus information provider) and how it is served.

Step 6: link the real and virtual communities

The final step in creating a healthy, thriving virtual community is to weave it back into the real community. Examples of successful integration include URLs on business cards, departmental e-mail addresses in the town newsletter, and coupons that can be printed out from the Web and redeemed. Another idea is a program to acknowledge Certified BEV Businesses complete with a BEV sticker in the window of every local business with an e-mail address or Web page.

In the early years of the project, periodic BEV socials joined the real and the virtual as people met, face-to-face, for fellowship and fun. We held the socials at a local restaurant that had the town's only cyberbar. It cost the project nothing. Individuals paid for their own refreshments, enjoyed free snacks, and could use the public Internet computer to exchange ideas. We supplied name tags (usually self-stick diskette labels), had a sign-in sheet, and used the digital camera to take pictures of the tables full of citizens. The pictures were posted on the Web's social registry along with some narrative and the names of the people in the photos. The pages helped people put faces to names and were a popular site on the Web. As more and more residents became connected, attendance at these socials dropped off, and they were eventually discontinued. Nonetheless, they were an important activity that helped develop community.

The community network gets increased visibility when online groups (e.g., listserv members) participate in community projects like clean-up days and Independence Day parades. That gives a concrete demonstration that the online people are just like everybody else: old, young, single, married, conservative, liberal, male, female, introverts, and extroverts.

We provide e-mail addresses to all governmental departments to facilitate communication between citizens and their leaders. The town government serves as a good example to local businesses.

We support community groups and civic groups with Web pages, e-mail accounts, and listservs. This communication medium often replaces telephone trees and paper mailings. These groups see their meeting attendance improve because members are better informed of activities.

The BEV supports having public-use Internet computers available in the schools, libraries, and other institutions. As people see the computers in the places in which they are comfortable, they are more likely to try it out.

Last, we like to share stories of how the network has affected the lives of our citizens. Every member has a great story to tell. In one case, the BEV and the Internet were used to inspire an eighth-grade girl who was interested in a career in marine biology. Living in a landlocked area, she had no role models or professionals to share her interest. With the help of her science teacher, she established an e-mail network with female marine biologists around the country. The teacher saw the student's self-esteem grow and her whole outlook on life uplifted.

The BEV Seniors listserv was an invaluable lifeline to a woman who was housebound with her Alzheimers-stricken mother. It also helped a member who was immobilized by hip surgery for a time. Meanwhile, a local real estate company has seen sales increase from an online home and land database that allows buyers to search for homes based on price, location, and other features.

Checklist for managing information in a community network

Each community is different, with unique needs and varying goals that influence the way in which the information for an electronic village would best be managed. The following checklist is merely a guideline for developers and a quick overview of the concepts covered in this chapter.

- Develop and distribute the tools:

  Establish an office or community gathering spot for the network.

- If you are distributing software, be sure it is robust and thoroughly tested.

  Provide adequate technical support in the form of:

  Manuals and documentation;

  Online updates and FAQs;

  Help desk;

  Consultants for house calls;

  Volunteer network.

- Identify project champions and invite members from all community groups:

  Businesses;

  Health professionals and organizations;

  Libraries and museums;

  Tourism and economic development groups;

  Educators (K–12 and higher education);

  Governmental bodies and agencies;

  Local clubs and organizations;

  Religious organizations;

  Artisans;

  Newspaper staff;

  Children, families, and senior citizens.

- Educate citizens in a variety of ways:

  Provide seminars and short courses in a computer lab to expose groups or citizens to the network.

  Offer tailored presentations to clubs and civic organizations.

  Give hands-on training to individuals.

  Select groups for targeted training (e.g., businesspeople and teachers).

  Deliver training through the community college, continuing education center, and libraries.

- Foster a rich information space:

  E-mail provides training and online directories.

  Listservs keep focus narrow, identify leaders, and keep members local.

  Newsletters are distributed electronically and/or on paper.

  Usenet News start with a small number of groups and let demand drive group structure.

- Deliver the message effectively:

  Emphasize local content and focus on the community's unique contributions.

  Strive for quality; quantity will come.

  Have a unified, nonpartisan Web site that is owned by the community.

  Encourage contributions, large and small; use Web forms for novice users.

  Include every sector of the community; diversity is important.

  Design for the user with minimal equipment and knowledge.

  Practice good interface design. There are many books available on this subject.

  Choose the right vehicle for the information, based on criticality and frequency of change.

- Link the real and virtual communities:

  Arrange socials, speakers, and meetings for online users to come together.

  Encourage businesses to merge new and traditional advertising media.

  Have online groups (e.g., listserv members) participate in community projects.

  Provide e-mail addresses to all governmental departments.

  Support community groups and civic groups with Web pages and listservs.

  Have Internet computers available in the schools, libraries, and institutions.

  Share stories of how the network has affected the lives of citizens.

12

Building an Online History Database

by Kenneth William Schmidt and
Andrew Michael Cohill

Overview

Over the past several years, hundreds of cities and towns around the world either have started or plan to start building community networks [1, 2]. This interest in electronic villages is due in part to the growing popularity of the Internet and the WWW and to the press coverage given to a few communities that are pioneering that type of network [3, 4].

Documenting the design of such projects could be of great use to other communities trying to build their own networks. Many of these newer networkers will look to the pioneers for guidance to learn from past successes and mistakes and to discover issues that may not have been considered in early implementations. In the beginning stages of their projects, these networkers may want to find strategies for soliciting funding from government agencies and local utilities and successful methods for promoting the project. In later stages, they may look for ways to educate the public about community networks and to encourage local businesses to participate.

Developers of other community networks will find such histories useful, but documenting what happens as a community network evolves could be of interest to its users as well, since it will contain historical information about community members, groups, and activities. For example, a parent may want to learn more about the events that led up to the first electronic PTA meeting, or a retiree may want to see who was instrumental in creating senior citizens' presence on the network.

What is the best way to create this documentation? By definition, a community network is a sprawling, distributed activity, with no one central authority figure. Furthermore, even if you tried to get project developers to document it, it is well known that designers are not very good at producing documentation; they think it is tedious and are not motivated to produce material that would only benefit someone else [5].

The HistoryBase project[1] seeks to find a viable way to document such a project in a collaborative and distributed way, addressing the tediousness and difficulties of documenting a large-scale system by involving the entire community. It leverages the fact that the network is part of the WWW, automating the collection of information and integrating collection of additional information with use of the community network itself. It was revised and rewritten in 1998 and 1999 to make better use of database to Web publishing tools that did not exist at the time of the original development.

What is history?

According to *Webster's Dictionary*, history is a chronological record of significant events ... often including an explanation of their causes [6]. Landes and Tilly define history as "the branch of inquiry that seeks to

---

1. This project was supported by National Science Foundation Grant CDA-9424506. The grant proposal, "Building a History of the Blacksburg Electronic Village," which was written in 1994 by John M. Carroll, Andrew Michael Cohill, Gary Lee Downey, Edward A. Fox, and Mary Beth Rosson.

arrive at an accurate account and valid understanding of the past" [7]. Vincent discusses the history of the notion of history itself. Primitive history was told as stories, with no questions asked about the theories upon which it was based, nor concerning the motives which actuated the narrator [8]. Vincent notes that the work of Herodotus is the first known example of "extended historical composition," in which Herodotus states that he wishes not only to record the events but also to give reasons for them as well.

A modern history is a record of a set of events along with an interpretation of those events. Our project helps to create a record of the events that have occurred in the BEV, so that others can use the record as a base for analysis of the BEV itself. The project also supports interpretation of documents or events through a document annotation system.

Historical databases

Much of the current research in historical databases has been in the area of design history and group or organizational memory [9–11]. This section briefly describes three systems—Answer Garden, Designer Assistant, and Design Intent—that attempt to decrease the effort involved in information seeking in software development organizations.

To find information using Answer Garden [9] or Designer Assistant [11], the system guides the user through a branching network of questions. The user continues until the question that needs answering is found, whereupon the answer is supplied. The response might include text or graphical images, or it may be an active node that performs a certain action at run time (e.g., querying a database). If the answer is not found or is incomplete, the user can use the system to ask a question, which is routed to the appropriate human expert. The expert then answers the question via e-mail, and either the expert or the original information seeker can put the answer back into the system. By this mechanism, Answer Gardens and Designer Assistants databases grow with system usage.

In the Design Intent [10] system, communication artifacts—project documents, release notes, meeting minutes, and so on—are captured and made available for project stakeholders to comment on, critique, or clarify. Collaboration is facilitated through two methods: annotations and bulletin boards. Annotations are a means of commenting, critiquing, and clarifying Design Intent documents. Bulletin boards are similar to newsgroups in that discussion threads can develop on a certain topic. While bulletin board topics need not be related directly to a particular document, they eventually may lead to the creation of a new document.

System architecture

For this project a historical database system (referred to in this document as the BEV HistoryBase or HistoryBase) has been developed for the BEV. This system records project-related documents such as e-mail messages, meeting minutes, press coverage, and newsgroup postings by members of the BEV. The HistoryBase system, while developed specifically for the BEV, can be easily extended to other projects or systems where recording historical data is of benefit.

The BEV HistoryBase

The BEV HistoryBase (http://history.bev.net) is a hypermedia history system that is used to integrate the collection, organization, interpretation, and dissemination of documents and other materials pertaining to the BEV. The system is accessed using the WWW and is linked to the other Web pages that make up the BEV Web site.

Collaboration in the HistoryBase

The HistoryBase is a collaborative authoring system but not at the document level. Collaboration in the HistoryBase means that a set of documents is created through the joint efforts of many individuals in the BEV project group and community. The HistoryBase does not directly support multiple authoring of a single document. The focus of the system is to support the collection and organization of existing documents, not to support document authoring. The authoring support is minimal and is focused on the annotation of documents that have been submitted by other users.

In the HistoryBase, each document can have only one contributor. Documents can be created in collaboration outside the HistoryBase and then submitted by one of the collaborators (or a third party) after the document is finished. A document in the HistoryBase cannot be edited or revised. To update a document, an author needs to submit a new version of the document as a separate object or add an annotation to the existing document. This design ensures the consistency between annotations to a document and the documents content. If a comment is made on a document, and that document is later changed, the comment may no longer make sense. This design also prevents history from being rewritten.

Data model

The basic unit of storage in the HistoryBase is a main document. A main document can be one of many different digital media types: text, audio, image, video, or other (Figure 12.1). A textual document can be one of many formats, from a project design document to a description of a BEV event. An audio or video document can be anything from welcome messages by BEV personnel to clips of press coverage of the BEV. Images can be real pictures or computer-generated graphics.

Meta-information about each main document is also stored in the HistoryBase (Figure 12.2). The title is stored separately so that it can be easily accessed and used as an identifier for that document. For documents that describe an event occurring over a period of time (e.g., installation of communications lines in an apartment complex), the begin and end dates mark the time period. If the document is not time-oriented, the begin date is used to represent when the document was written or released (e.g., BEV purpose statement, February 1994). When a document is contributed to the HistoryBase, the submit date is set by the system. The classifications are tags that the contributor can give

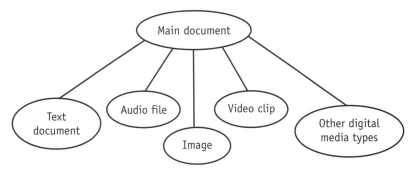

Figure 12.1   Main document data types.

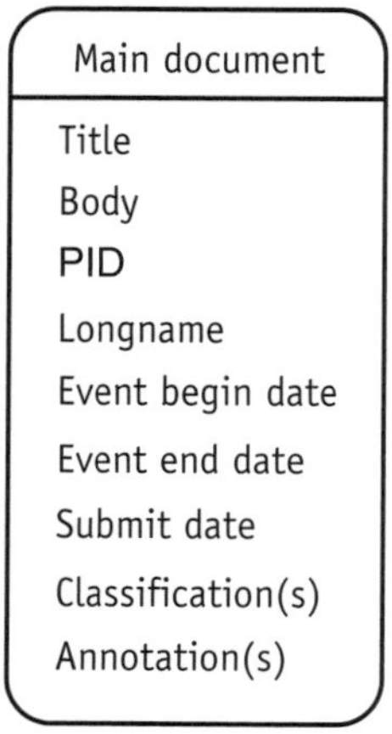

Figure 12.2   Main document metadata.

to a document when it is submitted. The PID and the Longname refer to the personal identifier (a unique code each person uses to access the network) and the real name of the person who contributed the document at the time it was submitted.

Finally, every main document in the HistoryBase is open to comment by any BEV user. The relationship between a main document and a comment is one to many; a single main document can have many different comments. Those comments are displayed below the main document in order from least recent to most recent. This ordering is used so that the flow of discussion can be seen from the original document through all the comments on that document.

Like main documents, every comment in the system has some meta-information stored with it (left portion of Figure 12.3). Each comment has a title or subject that summarizes the content of the comment. Information about who made the comment and the date the comment was submitted is stored also. Finally, there is a link from the comment to the main document to which the comment pertains (Comment On). Ordering of comments can be inferred from the submit date. For example, in the right portion of Figure 12.3, both comment X and comment Y are related to the main document, but since comment Y was submitted after comment X, it is displayed after comment X. This comment system allows many different perspectives on the same document to be shared among all users

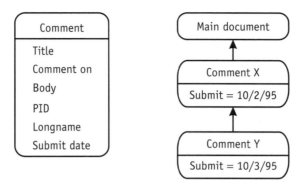

Figure 12.3 Comment subsystem.

of the HistoryBase. In that sense, the original document takes on a collaborative nature, with comments or reactions provided by any number of other users (including the original author).

Structure of the data space

All main documents, their comments, and the meta-data for each are stored in a database management system. This system allows flexibility in the type of structure that can be placed around the documents. Any attribute of these meta-data can be accessed and used as an index. For example, all documents in the HistoryBase can be organized by author, thus giving users the ability to study an individual contributors perspectives on the BEV.

Interface

Views of the data space

The main interface to the HistoryBase uses a timeline metaphor with landmark events (Figure 12.4). This metaphor is used because of the common use of timelines to organize historical events. The landmark events help to give anyone using the HistoryBase a bird's-eye view of the events that are a part of the history of the BEV.

Other views of the HistoryBase include a quarter-year view, which is a detailed view of the documents in a three-month period; a what's new view, which shows the recently contributed documents; and some

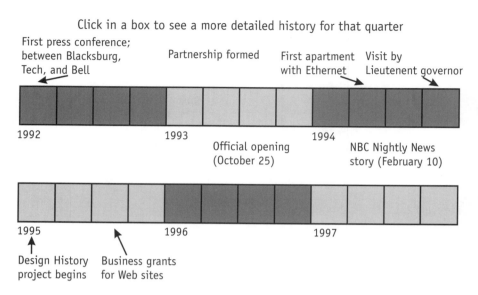

Figure 12.4   Timeline view of BEV history.

BEV-specific views of the HistoryBase. When the HistoryBase was rewritten in 1998, many more views and queries were added because of the power of the relational database used to store documents. These views of the system are discussed more thoroughly below.

Interface walkthrough

Any Web browser can serve as the interface to the HistoryBase (the Netscape [12] browser is used here for illustration). The basic element of the interface is a Web page; a collection of related Web pages is called a Web site. Figure 12.5 shows the first page a user sees when visiting the HistoryBase Web site. The title bar and the menu bar at the top of the page are parts of the Netscape browser, as well as the status section, which covers the bottom quarter-inch of the screen. These parts of the interface are not directly related to the HistoryBase and will not be discussed further here. The Web page itself starts with the HistoryBase graphic and extends down through the last updated line. Text that is underlined represents hyperlinks, which will bring up another Web page when clicked.

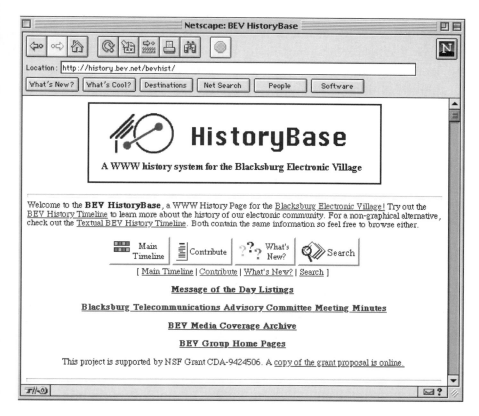

Figure 12.5 HistoryBase main page.

HistoryBase main page

The HistoryBase main page is the first page a user sees when visiting the HistoryBase Web site. The BEV logo is used in the graphic at the top to serve as a visual link between the BEV and the BEV HistoryBase. The four buttons in the middle of the page are listed as follows.

- Main timeline;
- Contribute;
- Whats new?;
- Search.

These buttons are hyperlinks to the main parts of the system (textual links are provided below these buttons for those users whose Web browsers do not support graphics). The four links starting with the Message of the Day Listing and ending with the BEV Group Home Pages are special sections of the HistoryBase that pertain specifically to the BEV.

Main timeline

The main timeline (Figure 12.6) is a high-level chronological view of the HistoryBase. Key events are labeled and positioned on the timeline; a user can retrieve a list of all the documents for a particular quarter-year by clicking on the corresponding section of the timeline. At the top and the bottom of this page, and every other page in the HistoryBase (except the main page), is a navigational button bar. This button bar is placed at

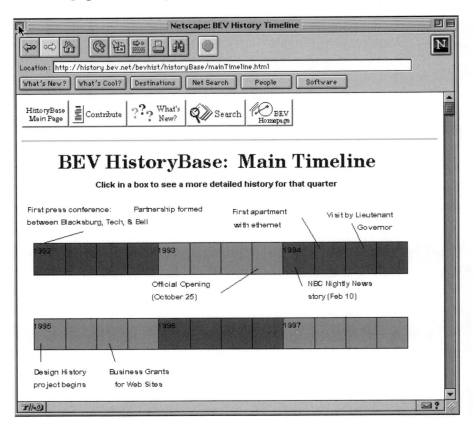

Figure 12.6   Main timeline. ©1997 Netscape Communications Corporation.

the top and the bottom because it is possible for the upper button bar to scroll off the screen when the user moves through a long page.

Quarter-year view page

The quarter-year view page is a document-level view of the HistoryBase for a three-month period. In Figure 12.7, the fourth quarter of 1995 is shown. At the top is the button bar, and just below that is the quick-click timeline. The prior quarter and next quarter buttons are useful for moving backward and forward through history; the quick-click timeline allows the user to jump from one quarter-year view to another non-contiguous quarter with a single mouse-click.

Below the quick-click timeline is the list of documents and events for the specified quarter-year. On the left side are the document dates and to the right are the document titles, which are also hyperlinks to the documents themselves.

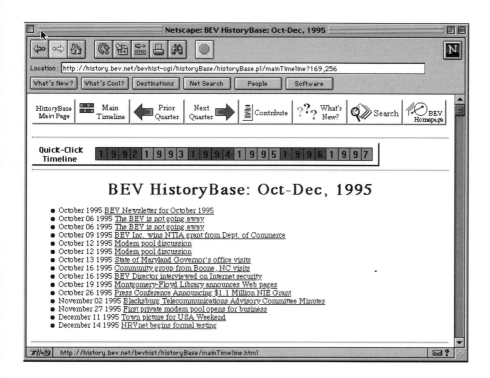

Figure 12.7 Quarter-year view page. ©1997 Netscape Communications Corporation.

Main document page

The main document page (Figure 12.8) is the display for a single document in the HistoryBase. A new button appears in the button bar on this page, the quarter year view button. This button is a hyperlink to the quarter-year view page in which the document resides. If the document is an event that spans over a quarter-year boundary, the quarter year view button links to the earlier of the two quarter-year views.

Below the button bar is the metadata for the document, followed by the body of the document. At the bottom is a link to the annotation page. The user can click on that link to make a comment about the document.

Annotation page

The annotation page is used to make a comment on an existing History-Base document. The body of the main document (and any previous

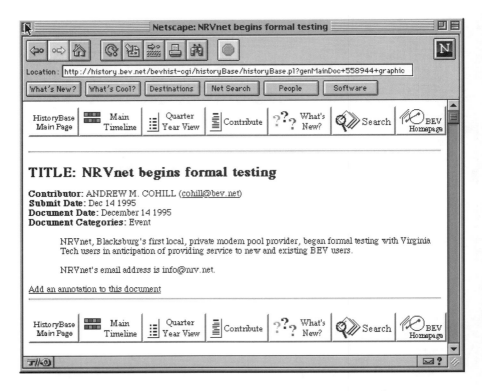

Figure 12.8   Main document page. ©1997 Netscape Communications Corporation.

annotations) is displayed at the top, and a text-entry box is at the bottom. When the comment is submitted, the main document is redisplayed with the newest annotation at the bottom. Figure 12.9 shows an annotated main document.

Contribution page

The contribution page (Figure 12.10) is used to contribute a new main document to the HistoryBase. At the top of the page is the contributor registration section. Anyone who contributes to the HistoryBase must enter a PID and password here. The system uses that information for authorization and to get the longname (see Figure 12.2) from the BEV authorization server.

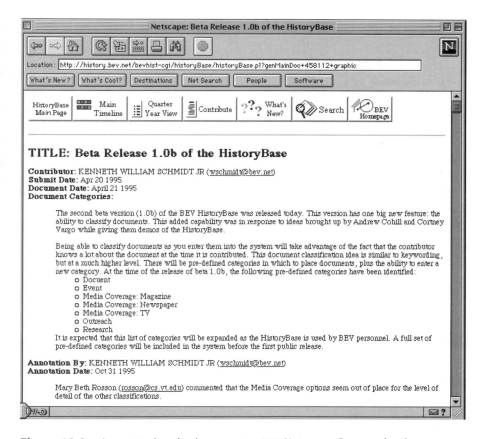

Figure 12.9   Annotated main document. 1997 Netscape Communications Corporation.

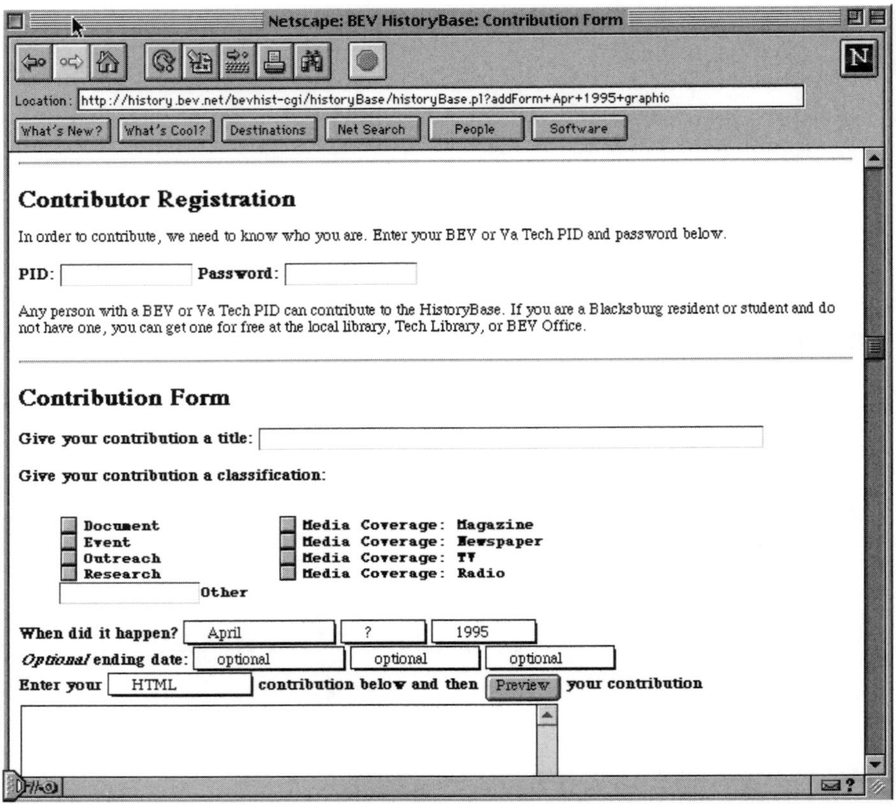

Figure 12.10   Contribution page. ©1997 Netscape Communications Corporation.

The next section is the contribution form, which is where the other main document information (detailed in Figure 12.2) is entered. The document title and classification are entered first, followed by the document or event date. If an event happens over a period of time, the ending date fields can be used to specify a time range. The body of the document can be entered in the text box at the bottom (in Figure 12.10, part of the text box has scrolled off the screen).

Other pages
One of the two other page types in the HistoryBase is the "what's new" page. This page is another time-oriented view of the document space. On

this page, documents are ordered by the date they were contributed, as opposed to the quarter-year view page, where documents are ordered by the document or event date. Clicking on the what's new button will always take the user to a list of the most recent contributions, whether those latest documents are clippings from a three-year-old newspaper or for an electronic town meeting held the evening before. The what's new page can be used by frequent HistoryBase users who just want to see a list of the latest documents available.

The search page is used to query for documents in the HistoryBase. Currently, searches can be performed on document titles, keywords (classifications), or in the body of a document. Results of searches are displayed in ranked order by the raw number of occurrences of the search terms in the documents.

Usage

The BEV HistoryBase is currently being used by several individuals and groups. It is being used as an online archive of the Blacksburg Telecommunications Advisory Committee (BTAC) meeting minutes. The BTAC has been involved with the development of the BEV since the early stages of the project and influences major project decisions. The HistoryBase is also being used as an archive of press coverage of the BEV. Not only do project personnel submit articles, but community members contribute news media coverage as well. Other current uses of the HistoryBase include documenting presentations about the BEV project and discussions about BEV-related activities. In 1999, the BEV group began publicizing the HistoryBase more widely to the community to promote its use by the citizens of Blacksburg. Lack of staff time, technical support staff, and training resources had prevented this earlier.

Issues

Implementation

The HistoryBase was developed on a DEC Alpha, running the NCSA WWW server. The programs that make up the HistoryBase are written in Perl and are accessed as common gateway interface (cgi) scripts through the Web server. Perl was used because its text processing features are more powerful and convenient than those provided by UNIX shell scripts or C. The back-end storage system is the Postgres database management system. All the Web pages in the HistoryBase are generated by Perl scripts, usually by generating a query to Postgres and then formatting the results as an HTML document. In the second version, Filemaker Pro, a relational database, was used to store documents and to provide the query processing for the system. Lasso, a Macintosh-based database-to-Web publishing tool, was used to provide the Web interface.

One of the obstacles to making the system widely available to the community was the reliance of the original version on the Virginia Tech PID (user id) authorization system. In 1998, the BEV group embarked on the development of a completely independent authentication (who you are) and authorization (what you are allowed to do) system that could be used by anyone in the community, regardless of their affiliation with Tech or when they became members of the BEV.

Use of the WWW
The WWW was used as the implementation platform because it is the least costly method to give HistoryBase access to the most people. A variety of Web browsers are freely available, and many of the potential users of the system already use the Web. Also, many of these same people view using the Web as something fun and interesting to do. It is hoped that using the HistoryBase also will be viewed as a fun activity and that people will enjoy making contributions to the system.

The pros and cons of anonymous contributions
The HistoryBase does not allow anonymous contributions. Several factors affected that decision, including the ability to determine validity and to assign a value to a document and the power of user identification in deterring system misuse.

If the HistoryBase allowed anonymous contributions, there would be no way to judge the validity of a document. For example, if the director of the BEV contributed a document on BEV personnel interactions, we would probably attribute more validity to that document than if the same document were contributed by someone not directly connected with the

BEV personnel. The same is true for assigning a value to a document. Knowing who contributed a document helps us to assign a value to that contribution.

Anonymous contributions could also lead to a degradation of the quality of the contents of the system because there would be no person connected to some documents the system contains. That lack of connection could open the system to misuse. A prankster might find it funny to post pornographic material, virally infected files, or other unacceptable material; a disgruntled community member might post an obscene letter to the mayor.

However, by not allowing anonymous contributions, opinions on certain issues may not be recorded. For example, an employee might have a completely different view of an event or policy than his or her employer stated publicly. The employee's job might be endangered if the employee publicly aired a contradictory view. If that information is not recorded, the database will not hold a complete picture of all the forces that influenced the history. Of course, it would be impossible to record all possible views of an event, but the lack of anonymity may cause us to miss some of the important views.

It is possible to permit pseudo-anonymous contributions by allowing a contributor to use a pen name that was issued by a trusted third party. With this method, no person or group, other than the original contributor and the pseudonym issuer, would know the identity of the author. This method could still lead to some document validity and value problems, but it would address the potential drawbacks of not allowing any type of anonymous contribution.

Contributor authentication

The HistoryBase authenticates contributors through a username and password scheme. This solution was chosen because all members of the BEV community already have usernames and passwords they use when accessing the BEV and the Internet. The HistoryBase makes calls to the BEV authentication server to validate usernames and passwords.

Several potential solutions were considered before deciding to use the username-password scheme. One possibility was to send the contributor an e-mail message whenever he or she made a contribution. If a user receives an e-mail message concerning a contribution made by someone else in his or her name, further action could be taken. This solution could cause problems for the HistoryBase system administrator. A single prankster could possibly submit tens or hundreds of documents on behalf of others. The HistoryBase system administrator would then need to remove all those false submissions.

Another authentication possibility is the use of digital signatures. A digital signature is a unique identifier that a person can use for digital transactions. This method is most attractive because it can be transparent to the user; however, the technology is not yet readily available. Another possible drawback of this approach is the export laws for some encryption technology used for digital signatures. If the history system will be an international one, this type of authentication may not be legally possible.

The HistoryBase uses the username-password method for authentication, but there are drawbacks to this method. One is that passwords can be snooped or cracked, thereby defeating the security of a username and password. Another is that people tend to forget their passwords, so password maintenance could become a burden on the HistoryBase system administrator. Finally, passwords can make the system seem less approachable to the lay user, which could be the most damaging drawback to the HistoryBase because it would tend to discourage contributions.

These issues are not important here because BEV users already have usernames and passwords and the HistoryBase can tap into the BEV user authentication system. The community-wide authentication/authorization system designed by the BEV group allows Blacksburg citizens to use the same username and password to access different kinds of information (e.g., the HistoryBase and the Village Mall). Users do have to be authorized individually for each use; that is, a HistoryBase user may not be able to edit a Village Mall business listing without prior approval. It may seem like a security flaw to use the same username and password in more than one place, but studies show that people tend to use only one or two usernames and passwords anyway. Studies also indicate that if users are asked to remember many passwords, they tend to create passwords that can be cracked or guessed easily. Using fewer passwords seems to encourage users to create difficult to crack passwords. The BEV group encourages the use of pass phrases, in which the password is created from the first letter of the words in a phrase (e.g., "The rain in Spain goes mainly to the plain" would be 'TriSgm2tp'). Pass phrases create passwords that are very difficult to guess or to crack. The username-password scheme may need to be considered more carefully in systems where that is not possible.

Future work

The HistoryBase is currently functioning as a part of the BEV Web site, but there are some improvements that can be made. Some of those suggested improvements are to various aspects of the system, while others suggest functionality that can be added to make the HistoryBase more flexible.

Adding more powerful search facilities

One improvement to the HistoryBase could be the addition of fuzzy searching capabilities. The current HistoryBase searching mechanism depends on an exact match between query terms and the terms in the documents. A search for the words clubs and groups will not return a document that contains references to community organizations. Fuzzy searching would require application of artificial intelligence techniques to information retrieval to improve the chances of finding the documents the searcher really wants. The use of a glossary in the search is one way of performing fuzzy searches.

Supporting multimedia contributions

Text documents are not the only type of history about the BEV that is being created. There are TV news stories, audiovisual presentations, photographs, and other multimedia aspects of the history of the BEV that were not fully implemented in the HistoryBase because of the limitations of the interaction techniques available in HTML: Text-based forms are supported, but audio and video file uploading is not (although Netscape recently released a new version of its Web browser, which has limited file uploading capabilities). The addition of multimedia object contribution will be a significant step forward for the HistoryBase. The second version of the HistoryBase has the capability of storing and retrieving graphics and other multimedia files, but training users to upload these potentially very large files continues to be difficult. This is largely an interface problem now, but remains one that is hard to solve easily.

Extending the automatic archiving capabilities

The HistoryBase has a few automatic archiving capabilities (like the newsgroup archiver), but a lot more information could be obtained this way. One possibility is to occasionally archive the top two levels of pages on the BEV Web Site. The HistoryBase already archives part of the highest level BEV Web page, where the messages of the day are listed. That is done every night, but the rest of the upper-level pages could be archived on a less frequent basis.

Another idea is to give an easy method for internal BEV e-mail messages to be archived in the HistoryBase. Perhaps BEV personnel could copy the HistoryBase on important e-mail messages. In the future, automatic archiving will probably be the main method for populating the HistoryBase. The difficulty with archiving e-mail automatically is protecting and maintaining the privacy of the individuals who are the senders or recipients of the e-mail. Certain kinds of e-mail can be archived easily (like public announcements to a mailing list), but many other kinds of e-mail are problematic.

Adding more powerful analysis tools

The goal for the HistoryBase project was to build a tool for collecting, organizing, interpreting, and disseminating documents and other materials pertaining to the BEV. Part of that goal that has yet to be designed is the ability to analyze the contents of the HistoryBase. A set of tools could be created that allows researchers to run complex queries on the data in the HistoryBase and to revise and refine those queries. The tools could also allow a person to leave behind the results of a query, even the query itself, for others to view and interpret and to use as the basis for further investigation.

Summary

This project has developed a historical database system that can be used for community networking projects and that can be extended to other systems in which the recording of historical data is of benefit. Although the HistoryBase is only starting to see use by members of the BEV community, usage should rise with a little more promotion. The HistoryBase of the future will have full multimedia features by which users can contribute sound, images, and video clips. It will also have a full suite of analysis tools so that the contents can more easily be explored and interpretations made. This project has taken the first big step toward an automated, self-documenting, hypermedia history system.

Acknowledgments

Mary Beth Rosson, John M. Carroll, Ed Fox, and Neil Kipp each made important contributions to the design of the BEV HistoryBase. The work was supported in part by a grant from the National Science Foundation.

References

[1] The USA CityLink Project, http://banzai.neosoft.com/citylink/, 1996.

[2] Morino Institute Directory of Public Access Networks, http://www.cais.com/morino/htdocs/pandintr.htm, 1996.

[3] Blacksburg Electronic Village, http://www.bev.net, 1996.

[4] Boulder Community Network, http://bcn.boulder.co.us/, 1996.

[5] Weiss, H. E., "Why We Still Have So Little Technical Documentation," *Infosystems,* Vol. 30, No. 5, 1983, pp. 88–89.

[6] *Webster's Tenth New Collegiate Dictionary,* Springfield, MA: Merriam-Webster, 1994, p. 550.

[7] Landes, *History as Social Science,* Englewood Cliffs, NJ: Prentice Hall, 1971, p. 5.

[8] Vincent, J. M., *Historical Research: An Outline of Theory and Practice,* New York: Lenox Hill Publishing, 1994.

[9] Ackerman, M. S., "Augmenting the Organizational Memory: A Field Study of Answer Garden," *Proc. ACM Conf. Computer Supported Cooperative Work,* 1994, pp. 243–52.

[10] Atwood, M. E., et al., *Facilitating Communication in Software Development,* Human-Computer Interaction Consortium, Snow Mountain Ranch, Frasier, CO, February 15–19, 1995.

[11] Terveen, L. G., P. G. Selfridge, and M. D. "Long, Living Design Memory: Framework, Implementation, Lessons Learned," *Human-Computer Interaction,* Vol. 10, No. 1, 1995, pp. 1–37.

[12] Netscape Communications Corporation, http://home.netscape.com/, 1996.

13

Success Factors of the Blacksburg Electronic Village

by Andrew Michael Cohill

PEOPLE OFTEN ASK, What has made a difference in the success of the BEV? Every community is different, geographically, socially, and economically, so I do not believe there is a single answer for all communities. The most important advice I have is to plan lightly. By that, I mean that too much planning can be just as damaging as too little. The technology and network services that flow from the implementation of technology are changing so rapidly that it is difficult to look ahead more than 12 months and know what might happen with any certainty.

This chapter discusses some of the most important things that we have learned in the BEV that seem to have made a difference. In brief, they are the following:

- Education rather than technology should be the focus of a community network.

- Show, do not tell, community members how to use the technology as a way of increasing use of the network in the community.

- Identify a project evangelist who can speak eloquently on the human use of this new communications tool.

- Public libraries are a primary focal point for community network education and use to ensure equality of access.

- Content drives community use of the network; breadth and depth are both important.

- Invest in an independent telecommunications infrastructure.

- Encourage public/private partnerships.

- Community support from all segments of the population is critical.

Education, not technology

Over the past six years, every problem that the BEV has encountered has turned out to be an education problem, not a technology problem. Indeed, if we failed at anything, it was understanding early in the project just how profoundly the question of education and training would affect everything we tried to do.

One of the assumptions that nearly all of us associated with the project had was that the idea of an electronic village was in and of itself such a compelling idea that the community would embrace this new communications medium quickly (the "field of dreams" idea—if we build it, they will come). We built it, and they did not come, not in the numbers we anticipated, anyway. Much work was required to develop the critical mass of users we needed to attract information providers—businesses and community groups that had services to offer and community information. The "field of dreams" model has been a stumbling block for many of the communities we have worked with—people change habits slowly, and a key role for the community network effort is to help people become comfortable with making changes in the way they communicate. Most people need some gentle assistance.

The business community was especially slow to get started with the idea of advertising on the Internet. It was not until the late summer of

1994, nearly a year after the official start of the project, that we offered a series of hands-on seminars for businesspeople in the community. We naively offered four 2-hour classes in a university computer lab; within a day of advertising them in the local paper, we had to develop rules to control who was allowed to attend, because we were swamped. Several businesses wanted to send as many as 15 people to a single class, and many wanted to bring four or five people. So we allowed only two people from each business, and filled all 60 seats within a week.

For many businesspeople, these classes were the first time they had ever seen the Internet; for nearly all of them, it was the first time they had the opportunity to use the WWW. For people who had struggled to master the intricacies of word processing and spreadsheet programs and the difficulties of Microsoft Windows, the one-mouse click interface of the Web was a revelation. A common remark after class was, "I never imagined it could be this easy."

After the first series of classes, we saw an immediate jump in the number of businesses using the Web in Blacksburg, and it has climbed steadily ever since. It was a revelation for me as well, because I finally understood the power of a little education. I had made a mistake common to technocrats, which is to assume people understand nearly as much as I do about something, but they seldom do. After that, we cut back sharply on presenting talks and demos to community groups and instead tried to bring anyone interested into the computer lab, where they could try the Web themselves. It was really the only thing that has ever worked.

The Internet is so different from the life experiences of most adults that I do not think it matters how much you tell them about it, it just does not mean much. They have no mental model for communicating instantaneously with hundreds, thousands, even millions of people. They not only have to see it, they have to try it themselves to believe it and to believe in it.

Demos do not work very well. By demo, I mean hauling an LCD panel and computer to a meeting room and using an overhead projector to show the computer display up on a screen for all to see. This visual display was always accompanied by a lot of hand-waving by me or someone else on the project while we exclaimed how wonderful the Internet was. It was never very effective because most people think they cannot use computers and believe that the demo looked easy only because one of the computer gurus was conducting it. They would watch with great interest but go away believing they could not possibly do it themselves because computers are complicated.

I began rethinking my entire approach to problem-solving on the project because of my experience with the businesspeople. When I really began to look hard at what we were trying to do, everything started looking like an education problem. I began to understand that we had a whole community of 36,000 people to educate, a few at a time. We had to educate businesspeople as to why it might be important for them to begin thinking of the Internet as a potentially important communications channel for advertising and for developing new relationships with their customers.

We had to educate local town and county government officials about the Internet and how it might be used to provide better services to citizens while lowering the cost of delivering those services—a rare win-win situation. We had to educate local community groups about why they should publicize their activities on the Net. When they did, they found that, among other effects, their attendance at physical meetings went up.

We had to educate consumers about the value of having the network connection in their homes, and how that connection (to the world) offered them better control over their time and their interactions with family, friends, and people in the community. We had to educate public school administrators and teachers about the value of having a network connection in the classroom. In every school I have ever visited, I have always found much tension about the real purpose of the network and the amount of additional work it might impose on already grossly overworked teachers. So we had to talk about possible shifts in the teacher's role in the classroom, and how the network might free teachers from some of the drudgery of delivering rote material and allow them to focus more on facilitating learning. And, of course, the way we did all this was to take teachers, 12 or 15 at a time, into one of our labs at the university and let them see for themselves.

In 1999, it may seem quaint to talk about teaching people the value of the Internet. The "dot com" hysteria in the stock markets and the apparently ubiquitous use of the WWW by business gives the impression that the job is mostly done. However, in early 1999, by most estimates, only about 35% of the country was online, meaning that the majority of people have not yet sent their first e-mail. Furthermore, we now have the advantage of hindsight in examining early efforts to get people online. In Blacksburg, even though about 75% of local businesses advertise on the WWW, many businesspeople are unhappy with the results. Some businesses have been very successful, but other business people still need more business and management training to understand how to integrate the Internet into their business.

Four years after the Internet became a household word in the United States, many school teachers are still inadequately trained and pre-pared to use the Internet effectively in classrooms. The focus in K–12 education has been to spend on infrastructure development (network connections, cabling, computers, etc.). While this is necessary, it is not sufficient. It would be better to slow down on spending (remember that prices drop by a third every 18 months) and allocate more money to ensure that when the computer and network connection in the classroom is turned on, the teachers have the training and the software required to use it effectively. Community networks can partner with schools to develop training programs for both teachers and the community at large.

Show, don't tell

This idea of pervasive education as a way of provoking change in a community leads to the important success factor of show, don't tell. The Internet is different enough from what people know that I believe that any successful community network project has to have a place to take people where they can use the network themselves to see firsthand what all the talk is about. In Blacksburg, because of our association with Virginia Polytechnic Institute and State University, we had access to a fully equipped computer lab with direct high-speed connections to the Internet, which is also very important. Blacksburg and Virginia Tech are particularly fortunate to have a New Media Center. The numer-ous New Media Centers around the country, developed specifically to support higher education and community use, are funded by several corporations, including Apple Computer.

Virginia Tech's New Media Center is a classroom equipped with 16 Macintosh computers for student use, an instructors workstation, and an LCD projector so students as a group can view demonstrations and software on a large display screen at the front of the room. The facility also has color scanners (essential for Web page design), a laser printer, and a variety of software ranging from basic word-processing software (ClarisWorks) to digital imaging programs (Adobe Photoshop, Fractal Design Painter) used for graphic design, as well as the complete array of software distributed by the BEV.

Most communities do not have university facilities like the New Media Center, so creating a similar local facility is extremely important. In Blacksburg, the local public library has played a major role in the education process; for many communities, the local public library may well be the most convenient location for a computer lab. The local library is also one of the first public facilities in any community that should get direct high-speed Internet connections, so putting the lab there may be less expensive with respect to the connectivity costs associated with running a lab.

Many localities are close to a community college, which is another place where the placement of a computer lab may pay off by doing double duty as a college teaching facility and a public use facility. The third place that may make sense is in a local public school, where again, double duty as a teaching facility for the schools and a public use facility during nonschool hours allow the community to maximize the use of the equipment and Internet connection.

In Blacksburg, citizens have also asked frequently for better access at the local town recreation center, which is another location that could be considered as a teaching and learning facility. In 1999, the BEV Seniors opened a computer lab in the town recreation center, with the assistance of the Town of Blacksburg and donated equipment from several organizations in the community. The BEV Seniors are training not only other seniors but are also providing computer training for children.

In the design of a computer lab, several different issues have to be resolved. The first is the space itself. Ideally, it should be large enough to support 12–15 town residents and computers, with a minimum of 3 ft of desk space for each computer, preferably 4 ft. One of the most common errors made in setting up such facilities is failure to provide enough open space on the desks for users to spread out workbooks, take notes, and work without bumping into their neighbors.

The classroom should also have room at the front for an instructor's machine and an overhead projector and color LCD projector. A screen for the projector is needed, as well as a white board for writing out information for the classes; a flip chart will do if it is not possible to mount a white board in the room.

The ideal configuration is to have one computer for each student and one computer for the instructor. However, if money is tight, two students could share each machine, so you could teach a class of 12 with just seven machines (six student machines and one instructor machine). Apple Macintosh computers are ideal for classroom instruction because they are easier to set up, easier to use, and easier to manage in a classroom environment. Macintosh computers are also superior multimedia and Internet computers. The Apple iMac is a superb classroom machine because of its one piece construction, built-in modem, and built-in Ethernet.

There is little difference between Macintosh and Windows network applications like Internet Explorer and Netscape Navigator, so Windows users will have little difficulty if they are trained on a Macintosh. Macintosh computers, with the addition of inexpensive software, can actually run Windows and the Mac operating system at the same time. Macs with the dual Mac-Windows capability are the most desirable lab machines because a mix of Windows and Macintosh users can be taught in the same class at the same time, with each student using the software with which he or she is most comfortable.

If your community lacks the money to purchase new equipment, used Macs are widely available for as little as $300–400. Eight-year-old Mac IIci computers run the latest Macintosh and Internet software and make excellent low-cost network computers. By comparison, eight-year-old Windows machines are not able to run the latest Internet software. Corporations and local businesses often replace desktop computers on a regular cycle, so it may be possible to solicit used equipment for a lab. If you are obtaining used or donated equipment, Macs are even more desirable because it is much more likely they will run Internet software and the latest operating system software.

Once you have identified a space, furnished it, and obtained computers, the last component in is the network connection. While it is possible to use dialup modem access (with SLIP or PPP) in a classroom situation, it is much better to obtain a direct Internet connection for the facility (see Chapter 10).

A T1 line is not the least expensive way of connecting a facility like this, but it is the best way. Other alternatives include xDSL, frame relay, and ISDN. Chapter 10 describes these technologies and how to use them. The direct feed is used to provide an Internet feed to the LAN that is often referred to as Ethernet. Check with your local ISPs about the availability of DSL services; in early 1999, DSL technology is the most affordable way to provide high-speed direct access to a computer lab.

The direct connection that feeds the Ethernet network in the classroom facility is important for two reasons. First, the direct connection to the network means that every computer in the classroom is connected to the Internet whenever the computers are turned on. That greatly simplifies teaching to a group because no one ever gets a busy signal dialing into the modem pool, no one ever gets a dropped line, and no one ever has any of the less common but irritating problems associated with modems. The teacher can concentrate on teaching people about using the network instead of fiddling with modem connections. Nonetheless, if it is not possible to obtain a direct connection, a classroom facility still will have much value even if modems have to be used.

Second, the high-speed direct connections allow members of the community to experience the Internet the way it was intended, which is fast. Modems, even the newer 56,000-bps models, are not nearly as fast as a computer connected to a local Ethernet and T1 connection. Eventually, everyone, everywhere, no matter where they live, will have direct connections. It will be easier to bring this kind of connectivity to your community if people can experience it firsthand. Remember that all problems are solved with education. Part of the education process is for people to see for themselves and use for themselves a direct connection.

The last thing needed for a computer lab facility is a supply of teachers. If you have been able to set up the facility, finding teachers should be easy. Local librarians, public school teachers, professors, students from a local community college, and professional computer users in the community are all candidates. In Blacksburg, we encouraged some of the local experts to offer classes through the YMCA. These classes are taught at the New Media Center, but the YMCA handles all the registration paperwork. It charges a small fee, but the fees guarantee that the instructors will continue to want to do the teaching work, and the classes are some of the most popular courses the YMCA has ever offered.

Find a project evangelist

You must find someone who is able to speak about telecommunications and networking in plain English. The quickest way to a slow start is to put a highly competent technocrat in place as the spokesperson for the project. As he or she burbles effusively in technospeak about the wonders of the network, potential users will quickly develop a glazed, dazed look in their eyes and begin to glance nervously toward the exit.

After hearing a talk about the network that focuses on technical minutiae rather than how it serves people, the audience will go away believing what they have always believed about computers—that they are complicated and hard to use. One of my favorite ways to start a demonstration of the Internet is to disconnect the keyboard from the computer and throw it in the trash can. I then hold up the mouse and ask how many people can push a button. Of course, everyone raises a hand, and off we go. After about 30 minutes of using the Web, I stop and remind everyone that the only input device we have used while traveling the world on the Web is the mouse, more specifically, exactly one button on that mouse.

This melodramatic demonstration of technology at its simplest is what it takes to make a community network a success. It is critical that people feel in control of this new technology. In all likelihood, their prior experiences with computers have been frustrating, and the prospect of trying this seemingly complicated new network thing layered on top of all the problems they remember just trying to get a printer to print out a three-page letter is frightening. Their first experience of the network, somewhat ironically, will probably be from the mouth of a real human being, and that particular human being should be quite comfortable talking with people.

Trying to manage (note that I use the word "trying") a community network was a great opportunity for me to grow as a person because I had to learn to be a much better listener than I ever had to be in any previous job. To paraphrase an old aphorism, you can lead people to the network, but you cannot make them drink deeply of the Internet. Many of our most successful efforts were the result, not of complex, inches-thick planning documents, but of ear-to-the-ground abrupt changes of plan, just because of what we heard from the community.

In Blacksburg and in meetings I have attended in other communities, I have almost always seen one or two self-styled computer experts in the room. These folks have two troublesome traits that can cause problems related to the explain the network in English challenge. First, they usually think they know more than they really do about computers and the network. That in and of itself is not necessarily a bad thing, but when you couple that with their second trait, it spells trouble in a room of computerphobes. The second trait they often have is a great desire to make sure that everyone understands how much they know about the network.

So in a meeting to answer questions in basic English about how the network might be a useful thing for the community, you have these folks lobbing the equivalent of hand grenades in the form of complicated technical questions. If you (as the speaker) get sidetracked answering too many of these usually irrelevant questions, people will (1) get bored and (2) go away believing the network is as complicated as they thought it was.

A subsidiary danger that I have also seen in some communities is that one of these know-it-all people (call him Bill) ends up volunteering to help get the network going but really only succeeds in confusing everyone around him and actually slowing things down. The trick is to spot Bill early, cut him out of the herd, and find something useful for him to do (Bill is often a very nice person and full of energy) that makes good use of his talents without giving him the opportunity to scare the wits out of all your other users.

Some of the ways that Bill can play an important and useful role in a community network are to do the following:

- Make house calls to help people get their software working;

- Test and debug new software that you may want to distribute to the community;

- Run a community software site for shareware and freeware;

- Organize a volunteer group to help with telephone or e-mail technical support to residents who need help;

- Help community groups design and set up Web home pages;
- Help in the community network storefront or office.

The golden age of libraries

In Blacksburg, we have been most fortunate to have a public library that early on recognized the potential of the network. Forward-looking librarians at the MFRL (http://www.montgomery-floyd.lib.va.us/) applied for a small grant in the early days of the project. The money they received from that grant enabled them to purchase 10 new computers for the library. Network equipment donated by Xyplex and a T1 line donated by Bell Atlantic enabled the library to be fully networked with a directly connected high-speed local network by the end of 1993.

Since then, the library has provided free Internet access to the community, and anyone who wants a free e-mail account can obtain one at the library. Demand for the machines was so high that the library immediately had to initiate a sign-up system for the public machines. Library patrons could sign up for 30-minute sessions and were limited to two sessions per day. In the first year of this service, in a town of 36,000, the library logged over 20,000 user sessions.

Demand has been steady since then. The BEV has taken universal and affordable access as a serious challenge, and anyone in the community, regardless of their ability to purchase a computer or network access from their home, not only can use the Internet in the library but can also have an Internet e-mail address. No one can tell where you read your e-mail, so the ability to put an e-mail address on a resume or reply to an online job advertisement or ask for advice online and have a reply address gives the economically disadvantaged the same power as those with network access in their homes.

In Blacksburg, the library has also been a critical partner in providing education and training. Part of the initial grant the library received provided for a part-time librarian to assist Internet users and to teach classes in the library. The classes were enormously popular, and the biggest problem was offering enough of them. A six-week series of seminars typically was filled within three or four days of their being announced.

Many people ask if the library, as an institution, will be irrelevant once we all have network access in our homes. I do not believe so. The library will always have a role to play in the community as a learning resource center. The transition to a fully connected society will take many years, and until then the library will play an important role as a provider of network access for those that do not have connections in their homes for one reason or another.

One of the most frequent criticisms of the Internet is that there is a lot out there, but it is organized poorly and it is hard to find. What a wonderful job for librarians, who are experts at categorizing information and helping others to find information. The information management problems on the Internet are very old problems in a very new medium, and I suspect that library and information scientists will play a major role in helping to organize these new sources of information and in providing access to them.

It will also take many years for everyone to afford a computer in the home, and until then libraries can play an important role in providing free access to the community. There is simply not enough grant money in the world to provide everyone with a computer at home. Even though a computer in every home is the most desirable long-term goal, in the short term I believe it is better to support community access through libraries and other public facilities than to simply say that it is not fair that some people have computers at home and others do not, then do nothing. Andrew Carnegie understood this simple principal. He did not propose to buy everyone one book of their own; instead, he knew that the community would be better served by the creation of a public, shared resource for books, which we now know as the public library.

Libraries, as a shared resource supported by the community, will also be able to provide access to network services that even 20 years from now may still be uncommon in homes. For example, libraries will probably always have larger network feeds than individual homes and so will be able to provide faster and better access to certain kinds of information. In the short term, I believe libraries will quickly become community video-conferencing and meeting centers. Community groups, meeting at a library conference room set up for video over the network, can provide anyone in the community with network access from home with the ability to attend the meeting via the network.

We may also go to the library to attend meetings that are taking place in other parts of the world by reserving the library videoconferencing room and attending via the network. This type of use will be common before the end of the century. It is important to keep in mind that this use of the network facilitates communication with other human beings. A common criticism of videoconferencing is that it will diminish human contact. That is certainly possible, but if I have to choose between attending a meeting via the network (human contact, however limited by the network) or not attending at all (no human contact at all), I know that I will choose the former, not the latter. Other applications I believe we will see in libraries include public access GIS (graphical information systems) and virtual reality environments like CAVEs (http://www.cave.vt.edu/).

The library in Blacksburg, despite widespread use of the Internet at home in the community, remains a focal point of Internet use and activity. All the branches of the library in the two-county system now have high-speed direct access to the Internet and Internet access to the card catalog. Six years after the start of the BEV, the public library in Blacksburg is still bustling with activity and remains a vital part of the community.

Breadth and depth of content drive use

During the first year we were open, when we were signing people up more slowly than we had hoped, we persisted in believing that the problem was simply that we had not yet found the killer application that would be so compelling that people would drop whatever they were doing and rush down to the office to sign up for Internet access.

It never happened that way, and eventually it began to dawn on me that there was no killer application, no next great thing that would cause prospective users to beat a path to our door. Now, as I look back over six years of trying to increase use of the Net, what worked was trying to find out what was important to people, one by one, and then showing them how the Internet could be used to support that thing or activity of value to them.

No matter what people say they do on the Net, in some way it involves communicating with others in some way that was not possible before or at least was much more difficult. In Blacksburg, the seniors group uses the Internet to talk about cataract surgery and where the best senior citizen discounts are. One of the most popular topics in the newsgroups is where to eat in town. One of the most popular newsgroups created for a specific topic is bburg.auto.repairs, where people talk (bluntly) about where to get their cars fixed.

One of the most interesting things that has happened is how the community uses the BEV services like newsgroups and the BEV-NEWS mailing list to criticize the management of the project (including me). I am often asked if this public (and sometimes vitriolic) discussion of my work troubles me. While it is always frustrating to see people misunderstand what we are trying to do, it is hard not to take pleasure in seeing that the system has enough value that people use it to discuss what bothers them.

The community network offers the community a public forum open to all with network access to discuss those issues that are important to anyone in the community. This role used to be served by the front porch of the general store or, in earlier times, by the public well or the commons. After World War II, with the dramatic growth in suburbanization in the United States, we lost those common spaces, fragmented by new roads and highways. The Net is no substitute for face-to-face human contact, but the reality is that we have had precious little of that as a community for a long time.

While it is possible that the network will fragment us even more by allowing us to choose very selectively what we wish to read and with whom we wish to converse, it is also possible that the very same tech- nology can bring us closer together. It is too soon to tell what will happen; in all likelihood, both possibilities will occur.

The role of a community network, supported by the community and not by for-profit ISPs, is to create a space in cyberspace that people think of as the electronic community. There are various ways to do this, but the most common and currently the most important is the community Web site.

Only a community-wide Web site focused on inclusiveness can provide equal space for every organization and individual in the community, regardless of their business affiliation or of their opinions.

In Blacksburg, what we learned was that we had to encourage and support a mix of services. There is no one kind of service (e-mail, WWW, mailing lists, online conferences, Usenet) that will convince everyone to get connected, and there is no one kind of information (local government, civic, business, etc.) that will convince everyone to get connected. To reach critical mass in a community, a community network must cast a broad net (pun intended) and seek to draw in as many people as possible.

Some may be uncomfortable with this approach, because the Net invariably attracts people who cannot resist the sheer power of shouting out their opinions to the entire community. This shouting is an opportunity for learning to take place, and the Net is big enough for everyone. Some mistakenly believe there is not enough space on the Net for everyone and that controls must be put in place to manage the participation of some people and the expression of some opinions.

There is simply no precedent in human history for the ability of a single person to broadcast his or her ideas directly to hundreds, thousands, even millions of people from the comfort and safety of home, all for a few dollars a month for network access. Let me say it again, because I believe it is an important idea: The Net is big enough for everyone and big enough for every idea in the world. It is big enough for good ideas and for bad; there is plenty of space for crackpots and for geniuses.

One of the mistakes newcomers to the Net make is to misunderstand how it works. This is easy to do when most of us have grown up in a media space controlled by a few gatekeepers and where scarce bandwidth has been regulated by the government. Some look at the unabashed discussions and uncensored information on the Net and believe that it must be regulated. But the Net has infinite bandwidth, and it can grow continuously as long as people want to use it, which means regulation is meaningless. Regulation is helpful only when what you have to regulate is limited.

The Net is different from every previous mass medium because it is a fully interactive, two-way communications medium rather than a one-way broadcast medium. That means when you turn on the computer, you choose what you want to read and what you want to discover, rather than having an intermediary like a newspaper editor, a TV broadcaster, or a radio announcer decide for you. That is a fundamental shift in power back to individuals and away from large organizations.

Low-cost direct connections

In Blacksburg, Bell Atlantic began offering low-cost routed T1 lines soon after the start of the BEV project. A T1 line is simply a telephone company term for a specialized phone line that is capable of transmitting up to 1.5 Mbps (about a thousand times faster than ordinary modems). A routed T1 line means that back at the telephone switching office it is connected to the Internet. So when a routed T1 line is installed in a building, computers in the building can be connected directly to the Internet at speeds up to 100 times faster than by using an individual modem on each machine. (For more information, see Chapter 10.)

Even more important, direct connections are on 24 hours a day, seven days a week. There is no dialing, no busy signal at the other end, and many fewer technical problems than with modems. For any organization with more than a few computers, direct connections are often cheaper than using modems when you consider the cost of providing phone lines, modems, and individual network access accounts for each machine.

For example, a single T1 line to a high school could easily support several computers in two dozen classrooms in the school, making the per-computer connection cost very low (as little as $5–10 per month). The drawback is that the initial cost of installation and network equipment can range between $3,000 and $10,000 per site, depending on the number of computers to hook up.

Direct connections in schools are very desirable. Teachers have consistently told us that direct connections are a requirement for any significant change in the way classes are taught. That is not to say that computers connected by modem are not a valuable resource, but the value added is often due to a dedicated and persistent teacher who has invested a lot of time making it work.

A common misconception is that direct connections are not possible in a community unless fiber is installed. The T1 transmission system is an old and mature telephone technology that was designed for copper phone lines. Virtually all the routed T1 connections in Blacksburg travel at least part of the way over plain old copper phone lines, and the telephone cable plant in most communities is able to support T1 connectivity.

The other thing that surprises many people is that most communities, large and small, already have large amounts of fiber cable installed. Fiber has been used to carry voice telephone calls not only for long distances but within communities for many years. The availability of fiber should rarely be a consideration when a community tries to start a network.

The availability of low-cost routed T1 lines enabled us to hook up the schools and libraries quickly. The T1 connection at the public library allowed the hookup of 10 computers for public access. In the public schools, lines donated by Bell Atlantic and other financial assistance by Busch Entertainment Corporation allowed us to connect six schools to the network. Because the schools in Blacksburg are part of a countywide system, we hooked up three schools in Blacksburg and three schools in a rural part of the county, where it was unlikely they would have network access for a long time without the help of the BEV group at Virginia Tech.

In each area, an elementary school, a middle school, and a high school were connected. One of the things we learned after a year of use is that in the elementary schools the volume of traffic was much lower, where most computer use is supervised closely by the teachers because of the ages of the children.

In many areas, xDSL services (see Chapter 10 for more information) are now becoming available. DSL services make use of existing copper phone lines, like modems, but typically operate at 3–5 times faster speeds, and are usually much less expensive than the now outdated (and slower) ISDN circuits. DSL connections have sufficient bandwidth to provide access to several computers, and provide a very affordable alternative to T1 and frame relay circuits for computer labs and small business use.

It is difficult to overemphasize the importance of direct connections in a community network. The Internet was never intended to be accessed via modem, and the modem and ISDN access, while important as a bridge technology, simply will not support some critical uses of the Internet, especially in schools, libraries, and community learning centers.

Businesses that are serious about integrated use of the Internet into their overall plans quickly discover that it is nearly impossible to do so efficiently using modems. Business use of the network is a critical part of community networks, because they provide access to local services via the network. If businesses cannot do that efficiently, growth and use of the network in the community will be slower.

Another area where the direct connections played a pivotal role in the project was in connecting apartments directly to the Net. In Blacksburg, 10 apartment complexes and more than 900 apartments have not only a phone jack in the wall, but an Internet jack as well. With Ethernet cards (which are cheaper than modems) installed inside their computers, the residents of an Ethernet-connected apartment run a cable directly from jacks on the back of their computers straight into the wall and are thus connected to the Internet 24 hours a day, seven days a week, whenever they turn their computer on, at speeds 100 times faster than using a modem—and no busy signals.

The apartments have been very popular, and the two most frequent complaints I receive are, "I cannot find an apartment with Ethernet" and "I live in an apartment with Ethernet, and it is too slow in the evenings." The enthusiasm with which residents of Blacksburg have embraced these direct connections makes one question the assertion that technologies like ISDN and cable modems will be good enough. ISDN and cable modems, if priced properly, are certainly attractive alternatives to using SLIP, PPP,[1] and modems to access the network, but I believe that within five years, many private homes and apartments will have connections supporting 100 Mbps and will be complaining that they could use a little more bandwidth.

For communities serious about making the transition to a 21st-century information economy, direct connections and affordable Ethernet in office buildings is the most important first step. A statewide Virginia task force on increasing high-tech jobs in communities some distance from major metropolitan areas found two key factors that influenced a company's decision to move jobs. The first was the availability of skilled workers, and the second was the availability of high-quality "Internet-ready" office space. See Chapter 14 for more information on economic development.

---

1. PPP, the successor to SLIP, is more robust and versatile than SLIP. It will eventually replace SLIP entirely.

Encourage public/private partnerships

When the BEV began, Virginia Tech was the only one of the three partners that had the expertise to provide dialup access to the Internet via SLIP. So the university, by necessity rather than by design, volunteered to offer SLIP access to the citizens of Blacksburg. Fortunately, the technology to support SLIP, PPP, and modems has matured, and community networks no longer need the services of a major research university to provide that service. In fact, I would strongly discourage any community network from trying to provide SLIP/PPP service unless unusual circumstances prevent a community from utilizing the private sector. In southwestern Virginia in August of 1994, Blacksburg was the only community south or west of Roanoke with any kind of Internet access. Six months later, nearly every one of the 15 counties in the area had one or more private IAPs providing local dial access to the Internet. In 1999, Blacksburg had more than a dozen ISPs offering a wide range of services at a variety of price points, starting as low as $10/month for limited service.

Dialup access is often referred to as a modem pool. Connection to the Internet via a telephone line requires two modems. Each user must have a modem connected to his or her computer and a phone line. Connecting to the Internet, or starting a SLIP session, as it is sometimes called, has several steps. First, the user runs a little computer program on his or her computer that calls, via the modem and telephone line, the ISP. The ISP has a bank of modems, called a modem pool, one of which answers the incoming call from the user's modem.

Once a modem in the modem pool answers the phone, the two modems talk to each other (a high pitched screeching sound), which starts a SLIP session, or a connection to the Internet. In addition to a pool of modems, other equipment is also required, which makes running a modem pool one of the most capital-intensive parts of a network.

Because modems and SLIP or PPP are trying to do what amounts to an unnatural act (connecting to the Internet via telephone), users often have problems getting SLIP to work properly. That means the modem pool provider must have several people (or more) providing technical support to help people with problems. In Blacksburg, users in the apartment complexes with direct connections call for technical support much less frequently than do users with modems. The cost of technical support for modem access can be more costly over the long term than the cost of the equipment.

The high cost, both in the initial equipment investment and the ongoing cost of technical support, means that dialup providers must have a revenue stream adequate to cover those costs. Add in the confounding factor that networking equipment often has a useful life of only about two years (because of changes in technology, not because it breaks down), and it means that a modem pool service provider must have a flexible rate structure and a very efficient organization to provide good service and remain solvent.

That is not to say that a community or local government should not undertake the provision of a modem pool at all. Rather, there should be an organized and extensive effort to explore for-profit support for that service first.

Looking ahead, it also seems clear that the now common dialup modem service may not be the way most citizens will be connecting to the Internet five years from now. Modem speeds are not likely to increase much farther than the current 56-Kbps standard. Newer, faster technologies like cable modems, xDSL services, and wireless services will provide better, more reliable service at comparable prices. DSL uses existing copper telephone lines but is capable of much higher speeds.

In rural areas, cable modem access may not be common for many years because most cable TV systems in the country are not "Internet-ready" and must be replaced completely to provide two-way Internet access. Rural communities will be the last to have cable TV systems upgraded. Similarly, DSL services may be offered late (or never) to many rural communities because of the small number of users and the longer distances between users and the telephone switch office. (DSL services work best when users are no more than three miles from the telephone switch.) If rural communities want a modern telecommunications infrastructure, they will have to build it themselves, in partnership with local entrepreneurs and local ISPs.

To the maximum extent possible, community networks should encourage private-sector investment as a complement to public investment and encourage public and private partners to cooperate in the development of a 21st-century infrastructure. There are roles for both in every community.

Community support

It may be stating the obvious to call community support a success factor, but there are several key players in the community that should be willing to work cooperatively on a community network. Those players include the local government, the local public libraries, the public school system, and some of the key business people in the private sector. If the area has a community college, a private or public four-year college, or even a university, the facilities and people of those institutions can become a valuable resource. Finally, an active group of citizens willing to help proselytize the concept of a new communications tool for the community is essential.

Unfortunately, in an era of shrinking budgets in every part of the economy, it is possible that some of these potential partners may view any kind of cooperative venture as possible competition for scarce funding opportunities. As always, education is the key to solving those problems. Although community networks are not a panacea for fundamentally difficult problems like poverty, lack of jobs, poor health care, and difficult economic situations, community networks do offer the community a new way of discussing solutions to those problems.

Ambitious local businesses may be able to find new outlets for their goods and services by using the Internet as part of their advertising and customer service strategy. Local government officials may find it easier to open dialogues with people who ordinarily would not visit city hall; they may also find new ways of reducing costs and improving service by using the network within city government. Local schools can bring new educational opportunities to students and teachers.

Cost is always an issue, but the costs of implementing a modest network on a community-wide basis is much lower than installing a new water or sewer system, lower than the cost of a new ladder truck for the fire department, lower than the cost of a community swimming pool, and lower than the cost of a new recreation center. These things are considered important, even critical, resources even for small rural communities. If the private sector takes part in helping to create a community network, through donations or preferably through business investment, achieving the goal of a community network is even more likely.

Community, not technology

In the development of a community network, the emphasis should always be on the community, not on the network. A network that does not serve the community well is not likely to be successful. When planning the physical network, the information structure of the network, and the network services, technical planners must continually pose the question, how does this serve the community?

It is no accident that the words community and communication have the same root. A community forms because a group of people with shared interests want to create a place in which they can share ideas, commerce, and common values. Cyberspace is no different from physical space in that respect.

I believe that a community network has the potential not only to provide a new kind of community but also to strengthen the existing community—if people use it to that end. A community network does not automatically solve difficult social and community problems, but a community of people using a network to communicate may find it a powerful tool to organize people with similar interests. In that way, the network, as a communications tool, serves the community.

14

The Future of Community Networks

by Andrew Michael Cohill, Ph.D.

The four key roles of community networks

Community networks' roles will change over time, but enough experimentation has occurred in enough communities to be certain that communities need community-wide networks. Community networks' four areas of activity and influence in the community can be summarized as follows.

- Communities need public spaces in cyberspace; community networks create those spaces.

- Community networks can fulfill an important role providing education and training to citizens, businesses, and government officials.

- Community networks can help a community shift its economic development focus to the new information economy.

- Community networks can help communities develop a robust, independent telecommunications infrastructure that serves both public and private needs.

Public space in cyberspace

The primary and most important role of community networks is to provide public spaces in cyberspace. Communities routinely invest in public building (town halls, libraries, recreation centers, parks, etc.) in the physical world. We must have the same kind of investment in cyberspace. It is not possible to have a fair and unbiased discussion of a sensitive zoning issue in an online forum sponsored by the Acme Real Estate Development Company. Tensions will inevitably arise between commercial sponsors of Web sites and civic groups whose aims do not fit well with the corporate mission.

In the physical world, civic and volunteer groups rely heavily on the availability of public space for their meetings and activities. Communities need to provide the equivalent in cyberspace—Web sites, e-mail, mailing lists, conference systems, and other online information services create the "space" online that communities need to support the public activities of their citizens.

With a majority of Blacksburg residents now online, we have completed our initial goal of "getting everyone connected." However, we still have much work to do: In the long term, one of the most interesting challenges is to understand how to best serve civic and democratic uses of cyberspace. The Internet offers great potential to change (for the better, we hope) the way citizens and local government staff and officials interact and exchange information about local community issues. To do that well will require a new generation of information tools designed expressly for that purpose and much additional training and education of both citizens and government leaders.

Civic groups were among the first users of the Internet in Blacksburg, creating Web sites and using mailing lists to keep in touch with members. However, early results disappointed some civic group leaders who were expecting dramatic changes. Over time, though, as most of the community acquired access and e-mail, those changes have come. For example, when 80–90% of the members of a civic group have e-mail, it does become possible to discard the paper newsletter and distribute information via a mailing list and the Web site. Paper newsletters, because of the high cost of printing and mailing them, often comprise the single largest budget item for some civic groups.

Newsletters are frequently published sporadically because in addition to the cost, the group must rely on dedicated volunteers to handle the large tasks of formatting the newsletter, taking it to the printer, picking up heavy boxes of finished newsletters, putting address labels on them, and transporting them to the post office. A task that previously took several weeks to accomplish can now be handled easily in an hour or two—send the newsletter via e-mail and update the Web site.

As noted in previous chapters, civic groups have also reported that their attendance at physical meetings tends to go up once they begin using the Internet. We believe this is because it is much easier to deliver information in a timely manner—a newsletter might be sent out once a quarter and may well end up in a stack of papers and mail at home, read just once months before some meetings actually take place. By contrast, an e-mail reminder for the monthly meeting can be sent out just a day or two before the meeting, thereby helping people to remember to attend.

It is exciting to see this kind of dynamic taking place, but it did not happen overnight. It requires a commitment from the community and from individuals in the community to struggle with the inevitable frustrations of using this new medium and sticking with it long enough to see the results. One mother related to me that her life was changed when the Blacksburg High School Band parents' group began to use the Internet. Previously, phone trees and infrequent newsletters were used to keep parents up-to-date about band meetings, practices, and games. It required a major effort on her part to make sure that her daughter had rides and that they were in the right place at the right time to pick her up and to be aware of cancellations and time changes. The band Web site and mailing list now makes this nearly effortless for her.

One way the community network project can speed use of the Internet in the community is to provide special training and classes targeted to civic groups (which the BEV did not do). If a civic group is interested in using the Internet, it is helpful to offer to provide a class just for those members of the group, so that they can see the value of getting group information more quickly.

The list of items below outlines the kind of services communities and community networks should provide to citizens. Note that residents will still have to pay for access (how they get connected to the network); access is expensive to provide relative to information services, and access is best developed in partnership with the private sector. Note also that i am not arguing that the community network should be the only provider of these services. The private sector can and should offer equivalent services; neither the public nor the private sector should have exclusive rights to provide such services. In the physical world, no one would suggest that a single municipal swimming pool should meet all needs in the community (i.e., outlawing private swim clubs), nor would anyone suggest that it should be illegal for the community to build a municipal swimming pool because it "competes" with private swim clubs. Citizens should be free to choose, and the community should be generous enough to ensure that everyone in the community, including the economically disadvantaged, has the opportunity to "exist" in cyberspace. The kinds of services that communities and community networks should provide to citizens are described as follows.

- Each person in the community should have an electronic mailbox (an e-mail address) regardless of his or her ability to pay. E-mail is an inexpensive service to provide, and there is no reason not to provide this to citizens as a benefit of living in the community. Public schools must do this anyway for students, teachers, and staff. It is neither difficult nor expensive to have the community network provide this for everyone.

- A common e-mail address space for every person in the community provides a sense of belonging—of being part of the online community. This can be accomplished with a community mail-forwarding service that is part of the goal of providing e-mail to the entire community.

- Communities should provide a physical space for access, meetings, short courses, and activities related to the online community presence. Local libraries are a logical place to provide network access for those who do not have network access at home, and as a source of and access point for network-based information. Libraries that are asked to do this should receive additional funding for "Internet librarians" and technical staff needed to do this properly.

- Communities should provide a common authentication mechanism (validating a person's identity) that can be used equally by both public agencies and private businesses to facilitate electronic voting and referendums and electronic commerce and to simplify access to services.

- Communities should provide a public, online directory of e-mail and Web site addresses of all personal, nonprofit, community, and business entities.

- Communities should provide support for wide use of mailing lists to facilitate discussions on any and all topics of interest to the community, especially local government issues and public education, and to facilitate the work of civic groups.

- Communities should develop and support a WWW server as a community information publishing resource in cyberspace for local civic and government activities. It is important to note that this is an effort that requires time and money to do properly. It is all too common for the community to regard this as a volunteer activity. This is a critical community resource, and the community network should be provided adequate funding to do this well. The Web site is the cyberspace equivalent of the "welcome" sign at the edge of town. It should be a clean, well-lit place that is maintained regularly by professional staff. The aim of the community Web site is to become the default home page for everyone in the community (i.e., the "portal" for the community).

- Communities should provide local Usenet server and news groups to facilitate a "town commons" where people can meet to discuss issues of interest asynchronously and to facilitate discussion and local commerce. Usenet is a wonderful if underused resource for people with common interests to share ideas and help each other. It is an inexpensive service to provide.

- Communities should provide professional-quality online confer- ence facilities to support moderated asynchronous meetings and civic discussions. Chat is not adequate for community discourse. There are now reasonably priced but full-featured conference systems available that support public discussion of critical commu- nity issues.

- Communities should provide a community history database to help document and preserve an online, archival record of important community activities and events.

- Communities should ensure that there is a comprehensive and affordable Internet training and education effort for local citizens who wish to use the network to participate more fully in the life of the community. In fact, local governments have a special obligation to create, maintain, and preserve public government spaces in cyberspace. Everything should *not* be for sale in cyberspace. There is a cost to providing public spaces in cyberspace, just as there is a cost to providing courthouses, town halls, and other public buildings. Fortunately, compared to brick-and-mortar construc- tion projects, it is an inexpensive thing to do.

Local government has an obligation to make government information widely available online. The country is in the early stages of a transition from a paper-based government available only when government offices are open to a government model that makes government information and services available 24 hours a day, seven days a week. Furthermore, as the citizens of the community begin to use these new communications channels provided by the network to discuss important civic issues, com- munity, government, and political leaders need to be participating. If they are not, they will discover that the community is making decisions without them. If we are to provide public spaces in cyberspace, they must be supported broadly, in the same way that streets, schools, parks, and other public amenities are supported—by the whole community. If government leaders are not using the network, they can't possibly understand the need to allocate resources to support it.

Here are the activities that local and regional government entities should be pursuing:

■ Local governments have an obligation to staff and citizens alike to develop and maintain a comprehensive five-year technology strategic plan. It is all too common for local government leaders (both elected and appointed) to avoid this because they are "uncomfortable" with technology or "don't know anything" about computers. It is long past the time that citizens should tolerate this know-nothing attitude. The world has changed; we need leaders who are willing to change with it. Community network staff often have much to contribute to this process. Unfortunately, many local government technical staff lack the skills and experience to help government make the transition. Government has an obligation to provide proper training opportunities to its staff and provide the funds to upgrade network and computing resources.

■ Local governments should provide access to all elected officials and administrative departments via e-mail. Local governments should provide adequate support and training to government officials and staff to ensure that citizen inquiries can be handled routinely and efficiently. Government officials who claim they are "afraid" of getting too much e-mail are not fulfilling their responsibilities to the citizens who elected them. Can anyone imagine a government official proclaiming that he or she is "afraid" of receiving written letters? If e-mail makes it easier for citizens to let their opinions be known, then government has the obligation to provide officials with the resources for dealing with it. Officials may need assistance sorting and managing e-mail, but this is not likely to be the most serious problem facing government officials in most communities, and it is certainly one that is eminently solvable.

■ Local governments should provide full access to all government information via a government WWW site. Local elected officials should ensure that there is adequate training and support to ensure that all paper-based government information can be routinely posted online as well. There is no longer any good excuse for not having a full-service government presence online; citizens should be able to fill out and send most forms online (this is why a common authentication system is critical) and to pay for most government services online.

Education and training

Some Internet pundit once remarked that the Internet is "the world's largest adult-education project." This is the way we need to think about it. We are asking the entire adult population of the world to learn a whole new set of skills. Most people need help with this learning process. Community networks have a major role in the community developing short courses and seminars, delivering training, putting together "train-the-trainer" programs for local schools and other public agencies, and pro- viding training to the general public.

Community networks can also assist in coordinating training resources in the community. Public libraries, community colleges, for-profit training centers, and K–12 schools all have potential to provide training of various kinds to the community. No one group or organization will be able to provide all of the training needed in the community because of the varying needs of different groups (e.g., school teachers are generally not able to take training classes during the day; business people generally want training only during the day, and so on).

It does not seem likely that demand for training will abate any time in the near future. In Blacksburg, once people had mastered the basics of e-mail, they began clamoring for instruction on how to make better use of search engines. Most people in Blacksburg still need help configuring their Web browser to access Usenet. Unfortunately, the growing com-plexity of software is making it more difficult for people to attain mastery of these tools, creating additional demand for instruction. New tools undreamed of 10 years ago are also creating demand for training. For example, the BEV is planning a program for musicians interested in learning how to encode their music in the emerging MP3 music format that has begun to change the music industry.

Currently, the FCC is considering new rules for low-power radio systems, which offer many opportunities for public and private use. As one example, communities may be able to economically develop low-cost, universally accessible emergency-broadcast systems. Properly funded community networks with professional staff will have the time and expertise to assist the public in understanding new technology and learning how to use it.

It is not likely that this rate of technological change and innovation is going to slow down in our lifetime. If communities are going to adapt, change, and integrate these technologies into the fabric of the community, some group or organization in the community must play an ongoing, nonpartisan role in helping community organizations understand how to make the best use of these tools and systems.

Economic development

The Internet is a fundamentally changing business. Communities without a 21st-century telecommunications infrastructure and citizens trained to use it will be left behind. The old brick-and-mortar model of economic development requires diversification and must recognize that the information economy is demanding new kinds of services, new kinds of workers, new kinds of office and manufacturing space, and new kinds of training opportunities.

It is important to note that these changes take time. Many communities are disappointed by their community network efforts if they do not see change in the first few months of a CN initiative. It takes much longer than that. Typically, it takes about two years for a community network project to develop momentum and to begin to have a measurable effect on the community. New kinds of job opportunities can begin to occur much earlier than that, but the world has embarked on a transition that will take another 20–30 years to mature.

Jobs creation

Community network projects have a key role as a catalyst in reshaping worker training and retraining in the community. A statewide study funded by Virginia's Center for Innovative Technology (CIT) found that the availability of skilled workers was the number-one consideration for companies considering moving work out of highly urbanized areas and into smaller towns and communities.

Most communities have a ready supply of workers young and old willing to retrain and learn the new skills needed in the information economy. As noted in the previous section, community networks will play an important role in helping the community develop the capacity to teach and train all those workers.

Community networks also play an important ancillary function in jobs creation by diffusing technology widely in the community. When a high-tech business relocates in a community, it will need not just programmers and engineers; it will also want to hire receptionists that know how to use e-mail. Such companies will also need shipping and receiving clerks that know how to use Web browsers to check overnight package shipments. In addition, they will need bookkeepers and accountants familiar with Web-based commerce and human resources personnel that know how to use the Web to find the best qualified staff.

There will also be a need for more specialized training that will require the resources and staff of the local community colleges, local four-year institutions, and the vocational/technical programs of the local public schools. Community networks help train everyone in the community on the basics, without prejudice, without discrimination. Community network-based training opportunities offer everyone in the community a chance to participate in the new economy.

As more and more people in the community receive Internet training and begin using the Internet, some of those community members begin to incorporate the Internet into their businesses, creating demand for skilled workers (usually earning higher than average wages). These positions are often filled by local residents who have received some training from the community network. As these local workers leave their existing jobs for the new work opportunities, vacancies are created and someone else in the community gets an opportunity to move up.

Internet-ready office space

A second finding of the CIT study on attracting high-tech businesses to smaller communities was the availability of high-quality office space. This means that communities and regions focused on traditional economic development activities like constructing shell buildings designed for manufacturing must diversify that strategy and include shell office buildings in the mix.

Community networks again play an important role in assisting local business, government, and economic development leaders in understanding the characteristics of "Internet-ready" office space. In many communities, the only building that is Internet-ready may be the one that houses the offices of the community network. Community networks can provide "show-and-tell" tours of their facilities, introducing local businesspeople and developers to the concepts of Category-5 wiring, between-floor risers, Ethernet, communications equipment closets, and flexible office space.

It is beyond the scope of this chapter to provide a comprehensive guide to developing Internet-ready office facilities, but some of the characteristics of such buildings are described as follows:

- Flexibility in rearranging interior walls to provide easy expansion for growing companies;

- Ethernet cabling and low-cost Internet access offered as a building amenity;

- Business incubator-style shared services like mail rooms, copier services, and building management;

- Easy access to ceilings and walls to install network cables;

- Between-floor risers for network cables and cable trays in the ceilings of hallways;

- Equipment closets on every floor with adequate cooling and ventilation for network equipment;

- Well-landscaped exterior space with recreational facilities like picnic areas, barbecue pits, and volleyball courts;

- Walking paths and bikeways within office parks;

- Bikeway and walkway connector paths to other parts of the community.

Business management training and development

In observing business Web sites, I see three prominent trends:

- Of those businesses that advertise on the Web, some have clearly not invested much thought or money in the effort and must be very disappointed with the results. A business with no clearly articulated marketing plan will not be able to make effective use of the Internet.

- A second group has approached the Web as a new advertising medium (which it is) and has had modest success in making its business more accessible to customers, via e-mail, online coupons, and clearly written and designed Web sites. Businesses in this group at least have a business strategy that produces measurable results.

- A third group has decided to approach the Internet strategically; these businesspeople have rethought their businesses from the ground up, integrating the Internet tightly with their core business (i.e., using it as more than a new form of advertising). It is this last group that has enjoyed the greatest success.

Community networks can help the business community learn how to take best advantage of the Internet by providing training opportunities, focused business development, and technology. A community network could partner with a local small business institute or training organization to ensure that businesses not only get the technology training they need but also get the complementary management and marketing training that will make technology-related business ventures a success.

There is a special opportunity for community networks to provide training and assistance to very small business ventures (microbusinesses), home-based businesses, and part-time businesses. Widespread, low-cost Internet access in a community creates many new opportunities for citizens who are now freed from the very high fixed costs of starting a business (like renting retail or office space). For many new kinds of business ventures, a modest Web site may be all that is needed to get started (along with adequate training for the business owner). Community networks can provide training opportunities and partner with local economic developers and government leaders to create virtual business incubators to help these microbusinesses to develop to a point where they can afford the services of private-sector Web companies, which often don't see these microbusiness startups as profitable customers.

Telecommunications infrastructure development
Fiber

Every community should consider developing an independent telecommunications infrastructure in partnership with the private sector. Governments can spur dramatic increases in economic development, jobs creation, and high-tech businesses by providing a public network of dark fiber and leasing out fiber pairs on a first-come, first-serve basis to businesses. Partnerships with ISPs and members of the business community that will want to lease fiber will allow the government to amortize the investment in the dark fiber over 15–20 years while keeping short-term assets like electronic equipment (routers, hubs, switches, etc.) in the private sector, which is much better at amortizing short-term assets.

The fiber itself requires little maintenance, so the government does not have to add new work positions to the public payroll, and even the installation can be contracted out to the private sector. This kind of shared development is the best possible use of public funds, because it creates a level playing field in the community for new and existing businesses that need advanced telecommunications services but allows virtually all jobs to remain in the private sector, creating economic incentives to grow quickly and to create increased tax revenue (which can be reinvested in additional fiber as demand increases).

Unfortunately, I think we are in for a long, rough ride with respect to access. Much of the work done by community networks in the late 1980s and early 1990s is being undone by the voracious appetites of companies that see bundling as a way of controlling not only content but access.

Want dial tone? No problem, but you'll have to pay for long-distance and Internet service to get it. Want an alternative Internet provider? No problem, but you'll have to keep paying for the one you have because it's bundled with your cable TV service.

Basic phone service in Blacksburg now costs about $28/month—a huge increase from the mid 1980s when you could get basic dial tone for about $12/month. Much of the increase is due to taxes and fees; who can say with any certainty that "access fees" have done any good? What do you tell a poor rural family struggling to make ends meet when they find out their "phone" bill is now $93/month because the company has thought-fully bundled cable TV and Internet access into lifeline telephone service?

In the future, community networks may be providing low-cost dial tone access. Start pricing used telephone switches or begin thinking about how to turn IP networks into voice networks. Communities that wish to remain independent, communities that want its citizens and business-people to have choice in the telecommunications marketplace, and communities that do not want growth stifled by high-cost, single-supplier network services must act now to ensure that competition is available. If competition does not exist, then communities must consider developing an independent, citizen-owned telecommunications infrastructure (primarily fiber, not electronics) that can be leased by anyone in the community on a first-come, first-serve basis.

Multimedia service access point (MSAP)

Virginia Tech and the BEV are heavily involved in Internet2 development, and we have had a special interest in the needs of small and medium-sized communities. In 1999, the BEV and Virginia Tech began operating a multimedia service access point (MSAP) in Blacksburg. The MSAP allows local data traffic to remain in Blacksburg rather than being transported up and down the East Coast.

This occurs because in most communities with multiple ISPs, the national Internet backbone is used as the switch point between data carriers. When someone in Blacksburg using a BEV e-mail account sends e-mail to someone using a private ISP e-mail account, the data traffic, until very recently, left Blacksburg and typically went all the way to Washington D.C. (and sometimes Atlanta). When local ISPs connect to the MSAP, Blacksburg data packets stay in Blacksburg, lowering the cost of Internet access for everyone in the community.

Even more interesting opportunities abound when a high-capacity, ATM-based MSAP is in place. Local videoconferencing (point-to-point, point-to-multipoint) suddenly becomes affordable. Likewise, video on demand begins to look affordable. An MSAP coupled with low-cost access to a municipal fiber system can create dramatic business and technology development opportunities in a community.

Wireless initiatives

The cost of point-to-point wireless Ethernet (typically in the 500-MHz to 1-Hz frequency range) has dropped dramatically in recent years, and in 1999 you can purchase a complete two-station wireless Ethernet system for between $5,000 and $10,000. The lower cost systems have a range of 10–15 km, and the higher cost systems can transmit as far as 25–30 km. In areas where there is no existing high-bandwidth wired infrastructure or where the cost of that wired infrastructure is unreasonably high, these systems enable communities to quickly build a fast, flexible data network. In Blacksburg, the cost of a T1 landline is about $425/month. By using a wireless Ethernet system, you would pay for the equipment cost in just one year. These systems are easy to set up and install, although antenna placement can sometimes add additional costs. However, tall buildings, water towers, and existing antenna towers can often be successfully used.

In 1998, Virginia Tech became the first and only university in the world to purchase and hold wireless licenses in the LMDS spectrum. LMDS uses bandwidth in the very high 31-GHz frequency range, which means that LMDS is useful over short distances (3–12 km) but at extremely high capacity (2 Gbps are possible). LMDS offers much potential in rural areas where developing a wired infrastructure may be unusually costly and/or where very high bandwidth is needed at low cost. LMDSs are not restricted to data; because of the huge bandwidth available in the LMDS spectrum, LMDS community services could include voice telephony, Internet data, video, and paging services, among others.

Virginia Tech and the BEV have begun to work with communities across southwest Virginia to help them design regional telecommunications infrastructures that incorporate both wired and wireless technologies. Communities interested in developing telecommunications infrastructure plans should consider how the community or region is going to take advantage of the new infrastructure, what new or innovative projects they would undertake, what groups in the community would use the technology, what problems they would try to solve, what comprises their existing infrastructure, and who will purchase services, among other issues.

Specifically, some of the things any community should probably consider are described as follows.

- Commitments and support from key groups in the community, including local businesses, local government entities, civic groups, schools, hospitals, large business enterprises, and libraries: In other words, any group that might benefit from enhanced telecommunications services ought to be a partner in the project.

- Existing infrastructure may be utilized to save money. Existing towers could be important, as well as strategically located buildings. If a wireless venture is planned, site studies are required to determine whether a site is suitable.

- Existing telecommunications and economic development efforts may improve or enhance an infrastructure development project (e.g., an active community network, Internet activities by local governments, Internet access and use in schools and libraries, economic development projects aimed specifically at telecommunications, etc.).

- Potential for combining wired and wireless projects to create major increases in connectivity for an area: LMDS is not the entire answer, and fiber is not always the answer either. In the future, both technologies are likely to be important. Can communities identify unique or novel ways to leverage wireless technologies with existing wired infrastructure, or new ways to combine the two?

- Financial support from the community: Telecommunications infrastructure projects will be a combination of public and private investment; communities that can put together a financial package that includes investment and/or cost sharing from local government and some private partners to help with infrastructure development will be well-positioned. Private partners might include ISPs that want to expand connectivity possibilities, large companies that want better or faster data services (including voice, video, and Internet access), and local public institutions like municipal and county governments, libraries, and schools. This is not necessarily cash upfront (although that always helps). A community might get commitments from several medium and large customers that would be willing to purchase services once an independent telecommunications infrastructure is introduced into the community.

Four biggest challenges

Any community network initiative will face four great challenges stemming from education; infrastructure; privacy, security, and content; and funding. These challenges are described in the following subsections.

Education

Education is discussed at some length in this book, but the importance of providing proper education and training in the community, especially to political leaders, economic developers, and businesspeople, cannot be overemphasized. Successful community transformation means that the entire community must reach a consensus on a vision for the community of the future and commit to following that vision to its conclusion. This kind of transformation will simply not happen without a widespread, comprehensive commitment to education. Community network projects are 90% community and 10% network (technology). Do not be confused or confounded by people spouting technical jargon. If someone cannot explain technology to you in a clear and simple way, send that person away and look for someone else to advise you. Technology (and the people that implement and support it) should serve the community; the community should not be a slave to technology.

Infrastructure

Choice in the marketplace of telecommunications services is critical to the long-term health of communities. For much of the twentieth century, communities spent enormous sums of money on municipal water systems because they recognized that clean water was essential to the long-term health of the community—not just physical health but economic health as well. In the first half of the twenty-first century, communities must ensure that they can provide adequate telecommunications services. Fortunately, it will not be as costly as municipal water systems. The public and private sector must solve this problem together.

Regrettably, many large telecommunications companies are too focused on discouraging competition and limiting choice. This is a serious problem for communities, and if the marketplace does not respond, communities must develop independent, citizen-owned telecommunications systems.

Privacy, security, and content

The Internet is under continuous assault from both the private sector and government. Many businesses are inexcusably careless about personal information they are collecting from users (names, addresses, credit card numbers) and/or are deliberately reselling, combining, and mining this information to learn more about who you are, what you do, and what you buy. Every Web site, public and private, should have a clear policy statement about how it uses personal information, and the organization should stick to it.

Security problems appear in the news regularly. Although such stories are often exaggerated by the news media, there are people who engage in criminal behavior by trying to break into computers, either for the thrill of it or to deliberately try to steal information. Community information systems must be managed properly to minimize the risk of security problems. Very small communities may not want to try to run their own servers unless they have access to qualified security experts to help manage these problems (note that the Macintosh is considered to be very secure—much more secure than NT or Linux servers).

Communities have to consider how to deal with content issues like ensuring that children have access to age-appropriate material, determining community policy for information that may be considered offensive (e.g., sexual-oriented material and hate group material), and trying to balance freedom of access with community standards. Every time there is a shooting or some other violent incident, some political leader inevitably calls for regulation of the Internet. It is important to understand that this is impossible to do. The Internet is being used as a scapegoat for societal problems that we as a society would rather not consider too closely. No one has ever been killed by a book, and no one will ever be killed by a

Web site. We must look elsewhere for solutions to society's problems; again, regulating the Internet will not and cannot work.

Communities concerned about content issues must deal with the problem proactively, using acceptable use policies to describe appropriate behavior and to explain what kinds of information will be allowed on the community information site. Those concerned about sexually oriented material should make sure that parents understand that they, as parents, and not the community network, are ultimately responsible for supervising the after-school activities of their children. Parenting is best done by parents, not by committees.

Funding

Funding for community networks and telecommunications projects is a challenge now and will remain one for some time. These initiatives, to be successful, must have broad support from the public to succeed. Of the hundreds of community network projects already under way in the United States and abroad, the most successful ones have developed stable funding. Stable funding allows for the retention of key personnel. Furthermore, it allows the hiring and retention of staff with the proper technical, educational, and social skills and for continuity of development programs.

Resistance from the private sector continues to be troubling. Community networks that have enabled new entrepreneurial ventures in the community are finding themselves under fire from the very same companies once they become successful, with the business owners (often trained by the community network) complaining that the community network is now "competing" with them. Businesspeople put pressure on government officials, who, seeing tax dollars from the businesses and no tax revenue from the community network, may refuse to support the community network. This is why education is critical. If our political leaders do not understand the potential of the community network to lift all boats on a rising tide of technology, the community network effort may fail for lack of support.

It is beyond the scope of this book to cover all the funding possibilities in detail, but the most successful community networks are operating as nonprofit businesses, deriving income from a broad set of public and private support in the community. Successful community networks must find ways to mutually complement and support the efforts of private businesspeople, and private businesspeople must understand that the community network creates and sustains private investment opportunities in the community. I do not recommend creating a bureau of public information. Because of very serious First Amendment issues, governments must provide support to community network and infrastructure development from a distance, either by contracting for services on a businesslike basis and/or by providing block grants (unrestricted or for a specific purpose).

Getting started: a two-page plan

For those of you who have read the whole book and are now feeling overwhelmed, and for those of who you skipped to the last chapter in hopes of finding a few nuggets of wisdom, here is a short plan that any community can start quickly (it would take 12–18 months to complete). Details have been deliberately left out because there is no wrong way to do these activities, and many right ways. Each community may find different ways to implement and achieve the goals listed below.

- Estimate e-mail use in the community right now. Is it 10%? Commit to tripling it in 18 months. Provide e-mail accounts to residents, teachers, students, civic groups, and local leaders.

- Design and develop a "train-the-trainers" program to deliver short courses and seminars in five topic areas. Train 12 trainers in day-long training sessions. Trainers completing the course will be able to teach one or more of the following classes (with each class lasting two to three hours):

  - Introduction to the Internet;

  - Advanced Internet use;

  - Internet use in the K–12 classroom;

  - Internet use for civic groups;

  - Getting your business ready for e-commerce.

- Provide a community Web site with a community directory where individuals can list their e-mail addresses and have free personal home pages. The Web site will also provide a central directory for local businesses and a central directory for civic groups online.

- Provide personalized assistance to at least 25 civic groups, with training, Web space on the community server, and assistance developing Web sites. Develop and implement a regional online job bank where businesses with openings can list themselves online. Develop a complementary resume bank to allow citizens to advertise their work skills online. Train at least 40 businesses to use the job bank, and train at least 200 citizens to use the resume bank.

- Train at least 24 local leaders to use the Internet and provide personalized assistance in purchasing and setting up a computer if they need it.

- Host two online conferences with up to 100 citizens (two weeks each) to discuss community issues and to introduce the community to online discussion as a community tool.

- Identify 12 public and private institutions in the community spending at least $1,000/month on telecommunications services. Begin monthly meetings with representatives of each organization to explore ways to reduce costs by aggregating purchases and considering the development of a shared telecommunications infrastructure in the region.

Getting started: a one-page plan

For those of you who read the two-page plan and still feel overwhelmed, here is an even shorter plan that any community can start quickly (it would take less than three months to complete). Again, details have been deliberately left out because there is no wrong way to do these activities, and many right ways. Each community may find different ways to implement and achieve the goals listed below.

- Select a domain name for your community project (e.g., the domain name for Smithville Electronic Village might be sev.net).

Register that name with InterNIC (www.internic.net). This will cost $70 for the first two years.

- Once you have your domain name registered, lease a Web site from any local Web service company (this will give you a URL for your project, like http://www.sev.net). This should cost no more than $35/month. It may even be possible to have space donated, but be sure not to accept "free" space that is tied to advertising or that gives the contributor some special advantage.

- Find the very best professional Web designer you can hire (or find one willing to donate services) and get this individual to create an overall site design for you (including graphics, navigation, content structure). Make sure you design the site to be the community portal to the Internet.

- When the site design is complete, begin using local volunteers to create content, using the HTML templates created by your master designer.

- Advertise your site to the community. Use bumper stickers, billboards, public access TV programming, and word of mouth.

- Keep updating, changing, and adding new material to the site.

Summary

There was a time in history when communities provided very few of the things we now take for granted as proper roles for communities: fire and rescue squads, parks, recreation, clean water, sewer systems, garbage collection, and paved roads. Public libraries are just over a century old. Prior to that time, libraries were a luxury for the rich. The Internet offers communities new ways to solve problems. Should we deny ourselves the opportunity to test this new medium?

Despite the hype, despite the rhetoric, despite the uncertainty, can we dare not to try to make our communities better? Does anyone truly believe anymore that the Internet is a fad? When grandmothers go online and stay online, it is not a fad. We must cope, we must change, and we must grow. We have only two choices: We can learn to use this medium and put it to work for us, or the new medium will end up using us.

Resources

It is impossible to provide a comprehensive written list of online resources. The sites below all maintain links to other sites that will help you get started.

Blacksburg Electronic Village

The BEV site provides a planning guide in PDF form as well as a variety of other documents that may be useful if you are starting a community network project.

http://www.bev.net/project/evupstart/

Association for Community Networks

The AFCN maintains a very good list of current resources on a variety of topics related to community networking.

http://www.afcn.net/

Communities of the Future

The COTF site has a wealth of resources related to the development of 21st-century communities.

http://www.bev.net/cotf/

The New Democracy Center

The New Democracy Center (NDC) has a special focus on helping communities to integrate technology into the fabric of the community in a sustainable way. The NDC offers a wide variety of services to communities interested in community networks, information services design, and telecommunications infrastructure development. NDC members are included in NDC grant proposals, have access to NDC staff for consulting and design advice, and receive regular reports on technology and community development.

http://www.newdemocracy.org/

About the Authors

Erv Blythe is the vice president of information systems at Virginia Polytechnic Institute and State University and has been instrumental in the development of the Blacksburg Electronic Village (BEV) since the earliest days of the project. He has been responsible for several major telecommunications initiatives; most recently he has overseen the development of a statewide ATM network for educational and governmental use.

Phil "Theta" Bowden has been in data processing since 1967. During the last twenty years, he has served in various IS roles at Virginia Tech. In addition to directing the installation of the Virginia Tech communications infrastructure, he was also a co-P.I. for the development of the Virginia Education and Research Network (VERnet). He is currently managing the massive replacement of mainframe (IMS) administrative applications with Oracle, Web, and NC-based systems.

Dr. Andrew Michael Cohill is an information architect and the director of the BEV for Virginia Tech. He has an eclectic educational background in architecture, ergonomics, and computer science. His career in computing and networked information systems covers more than twenty-five years work in private industry, consulting, and academia. Cohill is an adjunct professor in the Department of Architecture at Virginia Tech and teaches in the Industrial Design program. He has published numerous papers, articles, and book chapters and has spoken widely on information design, the Internet, and community networks. Cohill is active professionally in the Environmental Design Research Association, the Association for Community Networks, and the Association for Computing Machinery. More information about Cohill can be found at http://www.bev.net/cohill/ or write to him at cohill@bev.net.

Roger W. Ehrich is a professor of both Computer Science and Electrical Engineering at Virginia Tech, where he has been since 1976. A graduate of the University of Rochester and Northwestern University, he has been engaged in research in image processing and human-computer interaction for many years. Since the creation of the BEV, he has focused most of his efforts on network-based technologies for K–12 education.

Dr. Robert C. Heterick, Jr., is the president of Educom and the chairman of the board of Blacksburg Electronic Village, Inc. He worked and taught at Virginia Tech for many years and served as the vice president for information systems at Tech during the start-up phase of the BEV.

Dr. Andrea L. Kavanaugh is the director of research for the BEV and research fellow with the Department of Communication Studies at Virginia Tech. A Cunningham Fellow and Fulbright scholar, she has published on communication systems and institutions in the United States and developing countries, particularly North Africa and the Middle East. She has taught at Hollins College and Virginia Tech. She holds an M.A. in International Administration; an M.A. in Communications from the Annenberg School, University of Pennsylvania; and a Ph.D. in Environmental Design and Planning from Virginia Tech. As BEV director of research, Dr. Kavanaugh works closely with academics and other researchers on a wide range of scholarly projects, from digital libraries to the use and impact of broadband networking. She is active professionally in the International Communications Association, the MacBride Roundtable, the Middle East Studies Association, and the American Institute of Maghrebi Studies.

Cortney V. Martin was the assistant director of the BEV and was with the project from its earliest days. She was responsible for much of the BEV's information collection and dissemination including designing and maintaining web pages, running listservs, creating the BEV newsletter, writing user guides, and developing and delivering training material to content providers. Ms. Martin has a B.S. in Industrial Engineering and an M.S. in Human Factors Engineering from Virginia Tech.

Melissa Matusevich has been a public educator for twenty-seven years and is an instructional coordinator for Montgomery County Public Schools, Virginia. She is currently a principal investigator for the federally funded field initated study "PCs for Families" and is a doctoral student at Virginia Tech.

> http://pixel.cs.vt.edu/melissa/melissa.html
> http://pixel.cs.vt.edu/edufis/index.html

Scott J. Patterson is an assistant professor of Communication Studies at Virginia Polytechnic Institute and State University (Virginia Tech) where he has been teaching since 1992 after earning his Ph.D. from Ohio State University. Scott's work focuses explicitly on individual preferences for different mediated communication channels such as the telephone. His current work explores the uses and impact of multimedia telecommunication systems on individuals and communities. Scott's research has been published in the *Journal of Media Economics,* the *Journal of Broadcasting and Electronic Media, Communication Research,* and *Media Information Australia.* He also edited *Collaborative Strategies for Developing Telecommunication Networks in Ohio.* Scott is an associate with the Center for Advanced Study in Telecommunications in Columbus, Ohio, and is a member of Virginia Tech's Cyberschool Teaching Project. Scott can be reached at spatters@vt.edu.

Kenneth William Schmidt, Jr., earned his bachelor's degree in Computer Science from Clemson University and his master's degree in Computer Science from Virginia Tech. In both his education and his career, he has specialized in human-computer interaction and information technologies.

Luke Ward is the technology manager for the BEV and has been a core participant in the project since 1993. He received a B.S. in Computer Science from Virginia Tech in 1977 and has subsequently worked with a wide variety of campus departments on projects ranging from large-scale database and workflow applications to the creation of a complete client-server e-mail system. Outside work, Luke enjoys coaching his 9-year-old daughter's soccer team, and on rare occasions he has actually been seen playing the fiddle. His e-mail address is lukew@bev.net.

Index

Successful Business Strategies Using Telecommunications Services, Martin F. Bartholomew

Telecommunications Department Management, Robert A. Gable

Telecommunications Deregulation, James Shaw

Understanding Modern Telecommunications and the Information Superhighway, John G. Nellist and Elliott M. Gilbert

Understanding Networking Technology: Concepts, Terms, and Trends, Second Edition, Mark Norris

Videoconferencing and Videotelephony: Technology and Standards, Second Edition, Richard Schaphorst

Visual Telephony, Edward A. Daly and Kathleen J. Hansell

Wide-Area Data Network Performance Engineering, Robert G. Cole and Ravi Ramaswamy

Winning Telco Customers Using Marketing Databases, Rob Mattison

World-Class Telecommunications Service Development, Ellen P. Ward

For further information on these and other Artech House titles, including previously considered out-of-print books now available through our In-Print-Forever® (IPF®) program, contact:

Artech House
685 Canton Street
Norwood, MA 02062
Phone: 781-769-9750
Fax: 781-769-6334
e-mail: artech@artechhouse.com

Artech House
46 Gillingham Street
London SW1V 1AH UK
Phone: +44 (0)20 7596-8750
Fax: +44 (0)20 7630-0166
e-mail: artech-uk@artechhouse.com

Find us on the World Wide Web at:
www.artechhouse.com